ENVIRONMENTAL SCIENCES LIBRARY

Environmental Decision Making

ENVIRONMENTAL SCIENCES LIBRARY

Environmental Decision Making

An Information Technology Approach

James K. Lein
Department of Geography
Ohio University
Athens, Ohio

Consulting Editor:
John Lemons
Department of Life Sciences
University of New England
Biddeford, Maine

b
Blackwell Science

Blackwell Science

EDITORIAL OFFICES:
Commerce Place, 350 Main Street, Malden, Massachusetts 02148, USA
Osney Mead, Oxford OX2 0El, England
25 John Street, London WClN 2BL, England
23 Ainslie Place, Edinburgh EH3 6AJ, Scotland
54 University Street, Carlton, Victoria 3053, Australia

OTHER EDITORIAL OFFICES:
Arnette Blackwell SA, 224, Boulevard Saint Germain, 75007 Paris, France
Blackwell Wissenschafts-Verlag GmbH Kurfürstendamm 57, 10707 Berlin, Germany
Zehetnergasse 6, A-1140 Vienna, Austria

DISTRIBUTORS:
USA
Blackwell Science, Inc.
Commerce Place
350 Main Street
Malden, Massachusetts 02148
(Telephone orders: 800-215-1000 or 617-388–8250; fax orders: 617-388-8270)

Canada
Copp Clark, Ltd.
2775 Matheson Blvd. East
Mississauga, Ontario
Canada, L4W 4P7
(Telephone orders: 800-263-4374 or 905-238-6074)

Australia
Blackwell Science Pty, Ltd.
54 University Street
Carlton, Victoria 3053
(Telephone orders: 03-9347-0300; fax orders 03-9349-3016)

Outside North America and Australia
Blackwell Science, Ltd.
c/o Marston Book Services, Ltd.
P.O. Box 269
Abingdon
Oxon OX14 4YN
England
(Telephone orders: 44-01235-465500; fax orders 44-01235-465555)

Acquisitions: Jane Humphreys
Production: Irene Herlihy
Manufacturing: Lisa Flanagan
Typeset by Best-set Typesetter Ltd., Hong Kong

Printed and bound by Braun-Brumfield, Inc.

Printed in the United States of America

97 98 99 00 5 4 3 2

Library of Congress Cataloging-in-Publication Data

Lein, James K.
Environmental decision making
: an information technology approach
/ James K. Lein
p. cm. -- (Environmental sciences library)
Includes bibliographical references and index.
ISBN 0-86542-466-7 (pc)
1. Environmental management--Decision making.
2. Information technology. I. Title. II. Series.
GE300.L45 1997 96-48895
363. 7'00685--DC21 CIP

To My Daughters, Adriana and Elena

Table of Contents

Preface

The "environment," which forms the central theme of this book, includes the natural and human expressions that form the landscapes of the places we inhabit. The study of the environment is the concern of several disciplines, including the one I have been trained in, geography. You may wonder why a geographer is not only exploring the topic of the environment but linking that topic to the business of decision making, for certainly that is a subject matter better left for psychologists to wrestle with, not geographers. There is a twofold answer to this question. First, when exploring the relationship between human beings and the places we call home, how we make decisions about the use of our environment greatly affects the quality of our lives and the integrity of the environments in which we live. Second, some time ago I realized that geography was more than simply places on maps and slides of faraway lands displayed on classroom walls. The idea that geography was, is, and can be a relevant problem-solving discipline, and that geographers can make a contribution to the larger scheme of things even if it seems beyond our realm, was radical. This basic idea continues to direct and shape my teaching and research, and is an idea that I hope is conveyed through the pages of this book in the coverage of topics far removed from traditional geography, among them information technology.

Nothing stands to reshape traditional geography more than the information technologies described in this text, and nothing offers the potential for geographers to join in and contribute their expertise to the study of the environment more than the application of these technologies. This is not to suggest that information technology is new to the discipline of geography or to the other subjects that participate in the questions related to the environment. It is not. Rather, there is a tendency to view technology as a curiosity—something that has promise but is beyond the fundamental scope of the discipline. Obviously, I disagree with this point of view, and I set out to demonstrate the applicability of these techniques and to show that they are not beyond the abilities of the geographer or any environmental professional to understand and use. Information technology has value if we put value into it. It is an exciting application area and a field too important to ignore.

Readership

This book is not designed specifically as a textbook, although it could be used in advanced undergraduate and graduate courses covering the general application of information technology in the spatial and environmental sciences. In such a course, this book could provide an introduction to information technology with the assumption that students have some prior knowledge of the basics. In my own teaching I have used this material as the core of a course that examines the role of information technology beyond database systems such as GIS. Very simply, I wrote this book with the aim of reaching environmental decision makers who shape and transform the environment with their policies—who have heard of geographic information systems, expert systems, and neural networks, but may not appreciate or understand their applicability. I also wrote this book for those of you who are currently using these technologies, to remind you of the bigger picture. It can be very easy to ignore the overriding purpose of an analysis and to focus instead on the technology, as if it were the end rather than the means. Sometimes we need to look up from the keyboard, look out the window, and remember that what we are doing concerns and affects living things.

Acknowledgements

In any written work the writer carries into the project all of his or her own experiences together with the influences of many individuals. First I would like to thank my mentors, Dr. Richard Ellefsen of San Jose State University and Dr. Milton E. Harvey of Kent State University, the former for showing me that geography was more than places on a map and that geographers can make a contribution, and the latter for demonstrating that geographers don't need maps to make those contributions reality, but that they do need to think and take chances. I would also like to thank the series editor, Dr. John Lemons, for his direction and support on this project, as well as the editorial staff at Blackwell Science. Lastly, I wish to thank my wife, Christine, for her support, time, and patience, and all the other things that wives provide and endure when their husbands set out to write a book, and Adriana and Elena, the joys of my life.

J.K.L.

Defining Technology's Role

Recently much has been written about technology and the information revolution that is changing our lives, how we work, and how we relate to one another. Paralleling these exciting descriptions of the future, however, are the dire warnings of impending environmental disaster. Questions concerning global environmental change, biodiversity loss, land degradation, population pressure, and the environmental costs of human technological progress challenge existing political institutions and present a very different picture of how our lives may change.

Although these two images of tomorrow's world appear in sharp contrast, the two themes, environment and technology, are inseparable, particularly when viewed relative to the cultural adaptations that have propelled human populations for well over a million years. Thus, while it may be convenient to connect technology to the environment in an adversarial way, there is another interpretation of the environment-technology link—another feature of the debate regarding technology and environmental change—that is all too frequently overlooked amidst the mounting evidence that defines the environmental problem. First is the simple realization that we are the technology, and secondly, the environmental problems attributed to human actions are ultimately the product of a decision we make regarding how a technology is applied. Therefore, part of the solution to the environmental issues identified above is to make better decisions about the environment, taking into consideration a wider view of the relationship between people and the places they inhabit.

Making environmental decisions, however, requires information and the means to use that information effectively, and herein lies a very familiar problem—one that is as old as our species. It is not the problem of environmental despair or technological wizardry, but a more fundamental problem of information, how to acquire it, how to use it to make decisions, and how to make better decisions. It is in this capacity that technology makes its own contribution to remedy the environmental problems that currently confront our global society. In the chapters that follow, this contribution will be examined.

The Information Challenge

Throughout history, and certainly since the arrival of the Industrial Revolution, a technology has evolved whose sole purpose has been to facilitate the gathering, production, storage, and dissemination of information. Indeed, from early hunter-gatherer societies to today's post-industrial age, we have always looked for ways to enhance our survival and comparative advantage by refining the tools we use and the methods we've developed to apply them. Information presentation and transmission have played central roles in that process. First with signal fires and clay tablets, then with printed books and maps, and now with the computer serving as the catalyst, the amount of information available to decision makers has grown substantially, although the innate human capability to process that information has remained essentially the same. Taking advantage of the available information has required new approaches to accelerate accessibility to information, process and analyze information, store and utilize information, and perform these tasks systematically and reliably.

Today, with improved access, information is more widely available and better able to meet a multitude of diverse needs. Yet, in many applications, more information is being created and stored than can possibly be analyzed or utilized effectively, and this has contributed to an information glut. Consider the following example from landscape planning or resource management to illustrate this point. At the beginning of this century, critical information about the resource potential, developmental suitability, or general characteristics of a given geographic area could not be accessed without extensive field surveying and mapping. That information was typically unavailable in any form. The data gathering and the analysis and documentation that followed could easily span several months to several years. As this century draws to a close, a single satellite can gather quantities of information over a matter of weeks for any location on Earth's surface at a rate that overloads the cognitive abilities of its users and saturates any decision-making process that tries to make use of it. This suggests that as human activity becomes more complex, information is generated in ever greater quantity and variety.

Specialized technologies will need to be developed to master this information and to keep pace with the methods and institutions that generate it. This has become the challenge of the "information age," and a technology has evolved whose purpose is to find ways to refine and simplify information, present it in new ways and new forms, and change how information is viewed and applied in the process of solving problems.

The ability to store and refine information and represent it in new and different forms introduces an information technology that has quickly become a powerful and nearly indispensable part of post-industrial society (Moffatt, 1990). Yet, as with any technology, it must be applied in the appropriate context in order for its benefits to be realized and used correctly for it to be effective. The technology that supports the storage and dissemination of information as defined by the available computer hardware and software systems is not unlike the fable "The Emperor's New Clothes." We look intently, attracted and sometimes seduced by something "new" that appears to offer potentials that we may not completely understand. We adopt this technology, excited by what we think it can do, only to discover much later that it falls short of our expectations. This point was well stated by Moffat (1990) who noted that while information technology is essential for rational environmental management, decision makers must not only have easy access to relevant information, but also must connect information technology to the decision-making process. Although such technology can contribute greatly to decision making, major methodological problems exist that must be resolved before information technology can be fully integrated into that process. It is important to realize that information is merely an aid to the decision-making process, and that while this is an important role, information technology is no substitute for clear, well-considered decisions. Nor is it an excuse for avoiding the ethical or political implications concerning the use of the environment (Wigan, 1987). This last point introduces a very different kind of challenge, a fourfold challenge to

- understand the limitations, potentials, and capabilities of information technology when applied to environmental problems,
- identify which technology is best suited to a particular type of decision problem,
- resolve the methodological issues that confound the widespread application of information technology, and
- demonstrate how information technology can be integrated into the process of making environmental decisions.

The principal aim of this book is to accept the challenge outlined above and provide an overview of methods to guide the utilization of information technology for the environmental professional. Specifically, this book examines the theory and application of a suite of complementary information technologies directed toward decision making and problem solving in the broadly defined area of environmental management. Although the role of computer-based technologies in environmental management is not new, recent developments and innovations have offered the potential to improve how analysts and policymakers solve problems and make decisions concerning the environment through the use of computers. Several of these innovations germane to the goals of environmental management are described in this chapter.

Within the past three decades, environmental professionals have witnessed the introduction of a staggering array of methods and techniques designed to help improve productivity and job performance. From an academic perspective we can explain this as the shift and convergence of one or more analytic paradigms, starting with the introduction of statistical and quantitative approaches in the environmental sciences and geosciences in the 1950's, trending through the world of modeling and simulation characterizing the 1960's and 1970's, and now arriving at the era of the database information system and artificial intelligence systems of the 1980's and 1990's. Like a needle weaving in and out of a complex fabric, these methods have likewise weaved in and out of the complex task of understanding environmental process. By contrast, technology and method can also be viewed as a collection of tools placed in the hands of the environmental manager. While this more pragmatic view may not be as elegant as the notion of a paradigm shift, it suggests that environmental professionals have at their disposal an elaborate workshop. When confronted with a specific problem, all that is needed is the right tool. Of course, the questions of which tools and when and how to use them are sometimes difficult to answer.

As the pace of innovation in computer technology increases, these questions become even more difficult to answer. Similarly, as the problems confronting environmental decision makers become more immediate and complex, the need for technologies to manage and simplify complexity also increases. There is a point at which machines and people meet and where solutions emerge through the cooperative efforts of decision makers and their computers. This book centers around that point of confluence where the machine's potential is integrated with the manager's needs, exemplifying the underlying belief that through a blending of method better environmental decisions are realized.

While it may be convenient to remain focused on the nature of technology for technology's sake, the success of a tool is not the tool itself, but how it is used. The underlying assumption throughout this book is that information technology affects the ways in which people make decisions. Decision making, therefore, plays a central role in understanding information technology, as does the nature of information that funnels through this process. Therefore, in order for us to begin to appreciate technology's role in managing and facilitating the flow of information to the decision maker, the properties that define information as both a concept and a physical quality must be addressed.

Information in Perspective

Very simply, every organism, every organization, relies on information to exist. Information is part of all human experience, and its acquisition and processing are fundamental aspects of life. On the basis of information plus prior experience the organism or organization reacts. Reaction often takes the form of a conscious decision followed by a behavior that reflects an attempt to change the environment in which one operates, or one's relation to that environment.

Therefore, when we describe an organism, a corporation, or a public agency, for example, the same four steps are always involved:

1. taking in information,
2. evaluating information,
3. making decisions, and
4. behaving.

Information, as a concept, can therefore take on a variety of meanings and be explored in contrasting ways relative to the role of information in influencing decision-making behavior. Several of those meanings more relevant to the question of environmental decision making are summarized by Debons, Horne, and Cronenweth (1988), and include:

Information as a Commodity—assumes an economic value that influences control and possession of information.

Information as Communication—initiates the exchange of data that transfers understanding and meaning.

Information as Facts—defines items of data that are devoid of context.

Information as Data—identifies the product of symbols organized according to established rules and conventions.

Information as Knowledge—suggests the intellectual capability to extrapolate beyond facts and data to draw conclusions.

As these definitions suggest, the concept of information can be applied to a wide range of cognitive states, define differing qualities, and assume contrasting functional roles. A technology dedicated to the management of information must therefore preserve critical aspects of these conditions if information is to remain useful to the decision maker and maintain its relevance as a resource to those who depend on it.

In language, terms such as data, information, and knowledge are often used interchangeably. This, of course, contributes to some confusion when these concepts are used in a scientific context. In a broader sense, data, information, knowledge, and wisdom are part of a continuum, where one quality blends into another as the result of actions taken on the preceding state, with no clear boundaries between them (Debons, Horne, and Cronenweth, 1988). Therefore, a technology that "manages" information must mirror aspects of that continuum, linking the purely data-driven processes of information access and flow to the higher cognitive activities that explain knowledge and define wisdom.

Debons, Horne, and Cronenweth (1988) present a detailed explanation of this transition from data to wisdom that they call the "knowledge spectrum." The spectrum begins with one event that defines a condition or change in the state of the world. This state or condition is represented symbolically and organized using simple rules to form data. We perceive this data when it stimulates one or more of our senses, and continual exposure to this stimuli leads to an awareness of the data. This state of consciousness suggests that the receiver of the stimuli has acquired information that can then be stored in memory or written down. In either case, the physical or cognitive representation of data coupled with awareness forms the fundamental explanation of information that is so critical to its practical application. However, as we shall see, that explanation takes us only so far.

Higher cognitive processes are required to move beyond information. These processes place meaning or provide understanding, which enables the decision maker to analyze situations and put the information as received into perspective. At this point, the decision maker can employ judgment given the situations, conditions, and facts that influence them. Based on these intellectual activities, awareness leads to knowledge, and, as with information, knowledge can be given physical and cognitive representation. According to this model, wisdom implies that knowledge, expressed in the form of human judgment and enveloped in a set of values that are the product of a social or cultural context, has been applied.

Accepting the premise that information is simply the cognitive state of awareness given representation in physical form, information technology must define more than a generic term used to describe the electronic storage, retrieval, and processing of data. Information technology, in a larger sense, must embody new approaches to problem solving and new methods of providing, acquiring, manipulating, and transmitting information that leads to the exercise of wisdom. From this perspective, information technology carries an important social and intellectual potential for opening up tomorrow's world for exploration by providing analytical and communicative capabilities that greatly expand the powers of explanation and prediction. When viewed in this way, information technology describes a method through which information is assembled, analyzed, interpreted, and utilized as a basis for an informed decision.

The link to decision making is critical in this context, particularly in decision areas that are complex, multi-faceted, and dynamic in nature. The more complex the decision environment, the greater the role of information technology as provider and facilitator of a construct through which decision analysis can be performed (Clarke, 1989). Yet, no single technology or method is universally applicable to the many varied situations encountered in environmental decision making. Consequently, a merger or synthesis of technologies that complements the level of variety surrounding the decision problem is one way of enhancing the analytical power of the decision maker.

A Synthesis of Methods

Perhaps the greatest problem in the application of information technology is determining where and how this technology fits into the process of making decisions in general, and environmental decisions in particular. While the promise of information technology is great, there is a gap between the promise and its realization in the form of better decisions. This point was well argued by Lagendorf (1985). In a discussion that critically reviewed the role of computers in urban and regional planning, Lagendorf characterized the connection between technology and the decision maker as a roller coaster ride, where new methods are introduced and widely touted though seldom critically evaluated. Since Lagendorf's examination of the topic, many of the issues raised have remained unresolved. The major issues identified by Lagendorf (1985) include concerns that decision makers:

1. do not understand the technology, its limitations, or its capabilities,
2. need change and that the technology often lacks the flexibility to respond to changing needs,
3. often cannot specify in advance what their information needs are, and
4. frequently employ judgment and other "soft" criteria that cannot be accommodated by the technology.

These four issues illustrate the need to understand better the nature of decision making, to place technology into that process, making certain that it is applicable to the range of problems under consideration, and to suggest a model that integrates the relevant information technologies and makes them available to the environmental decision maker. Achieving these three goals requires a synthesis of method that connects our present understanding of the decision-making process to the available technologies that handle information. And successful synthesis raises several questions, listed below, that have not been widely explored by the community of environmental professionals who seek to exploit information technology and maximize its applications to the decision problem.

- What is an environmental decision?
- How are environmental decisions made?
- What is the connection between problem solving and the decision process?
- How does information technology support this process?
- Which information technologies are relevant to the environmental decision maker?
- What do these technologies offer?
- How can their potential be realized?

Bringing together a complementary array of technologies tailored to environmental decision making requires a good conceptual model in order to avoid unnecessary confusion. The model must first recognize a need and then match an information technology capable of satisfying that need. This user-centered view begins with the general problem of data storage, processing, and retrieval; moves to the issues surrounding data manipulation, modeling, and the tasks that describe the process of transforming data into information; and culminates with the larger questions of knowledge, its representation in usable forms, and the procedures needed to store, retrieve, and apply knowledge toward the manipulation and transmission of data and information. The model, illustrated in Figure 1.1, describes a root association between the fundamental data and information access needs of the decision maker. The model also describes the intellectual branching that occurs from the basic requirements of storage, retrieval, query, and display. This branching defines more specialized demands on the modeling process linking knowledge of process and method to the database environment, rendering it more intelligent.

Because environmental decision making ultimately assumes a geographic expression, the decision maker is concerned not merely with data and information handling, but with the special demands placed on an information technology by data and information that defines uniquely geographic characteristics. This spatial, or geographic, dimension of data must be preserved and employed where appropriate, and the knowledge of spatial process and the representation of information and data as spatial phenomena must be available. Certain information technologies are instrumental to the problem of spatial data handling and provide critical capabilities for representing and manipulating data, information, and knowledge with an inherently geographic component. These technologies orbit around the nucleus of a data-, information-, and knowledge-driven approach to environmental decision making, illustrated in Figure 1.2. Here, the decision problem draws on the appropriate information technology, given both the nature of the problem under consideration and the characteris-

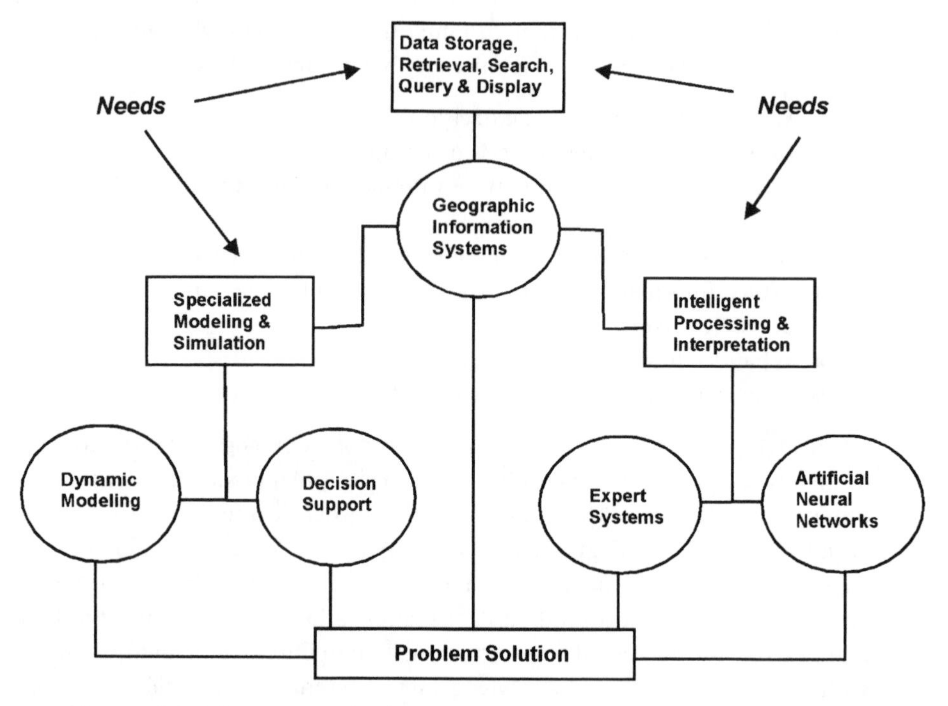

Figure 1.1 Information Technology System Design

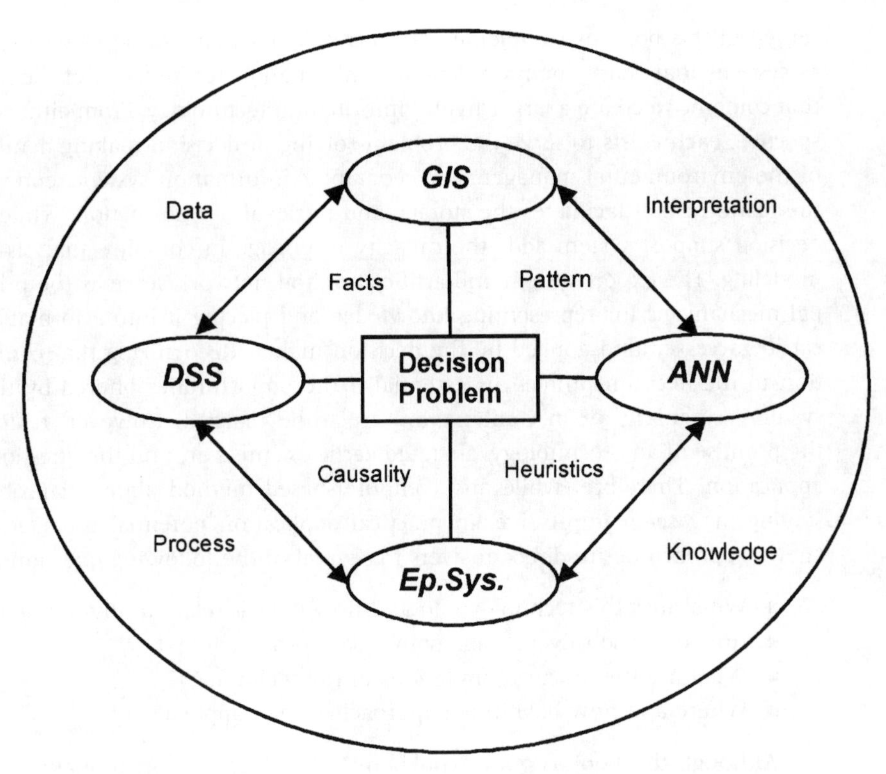

Figure 1.2 Information Technology—Environmental Decision-Making Schema

tics of data, information, and knowledge that define the problem and the intellectual tasks required to examine it. Included in this schema are the following information technologies together with their working definitions as used in this book:

- **Geographic Information Systems**—a computer hardware and software environment dedicated to the storage, retrieval, and analytic manipulation of spatial data.
- **Decision Support Systems**—a computer environment designed to support decision making by assisting in the organization of knowledge regarding ill-, semi-, or unstructured problems.
- **Knowledge-Based Expert Systems**—a computer program processing the capacity to perform reasoning operations using the knowledge of a human expert that has been coded into its structure.
- **Artificial Neural Networks**—a biologically inspired model that simulates the functionality and decision-making behavior exhibited by the human brain in a highly simplified way.

Although each of these four technologies can be examined as a separate subject area with unique properties and specialized applications, they are presented in this book as complementary technologies that can be viewed either as systems that share common data and information resources or as elements that combine to create a larger hybrid information technology. From either perspective, each exists to serve the problem-solving or decision-making demands of the environmental manager. The geographic information system represents the platform that facilitates the storage and retrieval of information, while the decision support system adds the capacity to engage in complex analysis and modeling. The expert system and artificial neural network serve as the principal mechanisms for representing knowledge and placing it into a format that can be accessed and applied by the decision maker. Recognizing that connection to the decision problem is essential if the opportunities offered by these systems, singularly or in combination, are to be realized. However, realizing the promise of any technology, as stated earlier, centers around the question of application. Therefore, while any computer-based method aimed at problem solving may seem impressive, in practical application potential users of any "new" approach desire direct answers to several of the following questions:

- What does this technology do and how does it relate to my problem?
- How do I use this technology in a day-to-day context?
- What are the practical limitations of the technology?
- Where and how have these approaches been applied?

Although this book is not a "cookbook" of recipes on how to make a geographic information system or an artificial neural network, each technology is placed in a problem-solving context, and its salient characteristics and essential features are explored and linked to the larger question of environmental decision making. Relating information technology to the decision process begins in Chapter 2 with a detailed discussion of decision making and the decision sciences. This overview of decision theory, viewed from an environmental perspective, provides background and identifies the main issues and factors that influence the design and utilization of computer-based decision aids. The chapters that follow present an applications-directed examination of those technologies with demonstrated capabilities to satisfy one or more requirements of complex environmental decision making. This treatment begins in Chapter 3 with an overview of geographic information systems and the issues that affect their use in the decision process and continues with a review of modeling and the concept of decision support in Chapter 4. The connection to knowledge in the decision-making process is described in Chapter 5 through an introduction to artificial intelligence and the fundamentals of knowledge-based expert systems. The role of knowledge is explored further in Chapter 6 within the concept of neurocomputing and the prospective benefits offered by artificial neural networks.

2

The Nature of Environmental Decision Making

Few topics have been the focus of as much concern as the environment. Concepts such as environmental change, environmental impact, and environmental health are examples of subject areas that have grown out of this concern, and they illustrate the myriad ways in which the environment has become a central theme in social, economic, and public policy decision making. These topical areas also reveal how complex the concept of the environment is and how almost every facet of human life and human activity can be explained in an environmental context. Whether expressed on a global or local scale, the environment is an ever-present factor that demands consideration. When a decision must be made that will affect the environment, deciding on the appropriate course of action given the interrelationships and interconnections that define the environmental system introduces an intractable level of uncertainty into what might at first glance appear to be a relatively straightforward task. Decisions are made, however, and although they affect the environment, the environmental ramifications of those decisions are typically considered as an afterthought. Yet these are the decisions that contribute to the degradation and destablization of our global environmental situation and lend support to the conclusion that this decision-making process is somehow flawed (Stern et al., 1992; Gallopin, 1991; Robinson, 1991). The question of how to change or redirect the process of making environmental decisions is of critical importance.

Within the past two decades several approaches have been introduced to better integrate environmental considerations into the decision process. Environmental impact assessment, environmental risk assessment, and environmental performance assessment are examples of developing methodologies designed specifically to bring environmental concerns to the attention of decision makers. Although the utility of these and other techniques has been demonstrated, any approach to integrate the environment into a decision process must contend with the problems of data, definition, and uncertainty. Therefore, inclusion of environmental information in a decision-making process depends not only on the ability of the decision maker to access, manage, and integrate data, but also on an understanding of the issues involved in making decisions and how data and information feed that process. In this chapter the decision-making process will be examined and the issues surrounding data and uncertainty will be explored.

The Environmental Question

It has been suggested that environmental decision making occurs whenever a decision must be made that affects the present or future quality of the environment (Chechile, 1991). This may seem obvious, but arriving at an appropriate environmental decision is anything but obvious, because the environment defies simple definition. At its most basic level, the environment is considered synonymous with the concept of an ecosystem, a set of elements connected by a flow of matter or energy that displays a discernable structure and occupies a definable spatial and temporal pattern. The set of elements selected to describe this vision is usually fairly narrow, and the focus of concern is generally exclusive of human activity. Although the ecosystem may serve as the basis of the environmental question, the environment explains more than the intricacies of an ecosystem and includes all natural and human processes that act on a clearly delineated geographic surface, and their union. Characterizing the environment in this way takes us back to the concept of the geosystem and employs this schema as the primary instrument to achieve landscape synthesis (Zonnefeld, 1983). Concern is no longer reserved exclusively for human or physical landscapes but rather their combination, recognizing that they interact. Applying the geosystem model to the question of human impact enhances basic understanding of the connection between the environment and the decision-making process. As suggested in Figure 2.1, human action, expressed in terms of land development, depends primarily on human perception of the components of the geosystem. Assuming that perception motivates action, an evaluation is made that ascribes a benefit to be derived from the geosystem based on how the system is perceived. The evaluation process, expressed in a highly stylized fashion, involves trading physical constraints for immediate cultural motivations. The decision to utilize the system emerges from this trade-off, which culminates in specific forms of use replacing the natural surface (Lein, 1986). In this way, the environment becomes an elastic concept that can be expressed differently according to problem-specific requirements. Thus, making a decision that can affect the future quality of the environment becomes first a question of conceptualization and secondly a problem of definition.

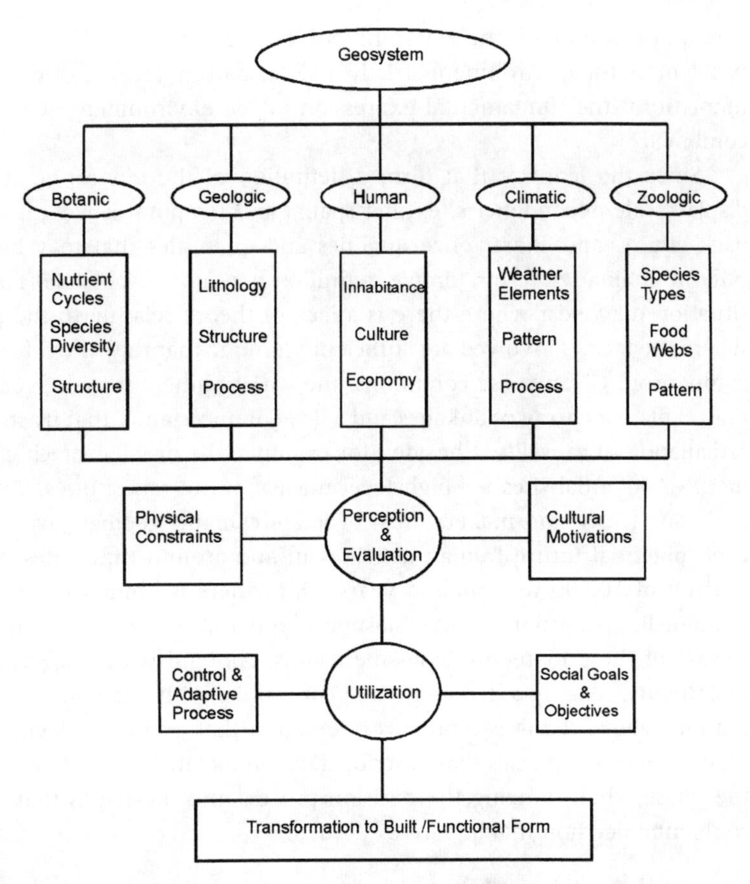

Figure 2.1 Generalized Components of the Geosystem

Taking the concept of the environment in its broadest sense reveals a multi-faceted, multi-dimensional complex characterized by direct and indirect interrelationships and interdependencies. Conceptualizing a problem in an environmental context must focus on questions of how the environment is represented, what elements compose its structure, how those elements relate, and what process or processes govern its behavior. Then, when a decision is implemented, focus shifts to consider how these effects are translated through this representation. To illustrate this point, consider the issues surrounding the siting of a solid waste landfill. Identifying possible sites requires consideration of surface and subsurface hydrology, geologic conditions, and ecosystem processes, as well as careful attention to existing land uses, property values, and land ownership patterns. Deciding on an appropriate location will in part be determined by how well the landfill is expected to perform given the environmental situation defined by these characteristics. Thus, conceptualization also demands careful definition of the environment as viewed from the context of the problem. Is the environment strictly a set of natural factors, or do human elements form the fundamental expression of the environment? Or are the two combined?

Since the features that form a definition of the environment will likely display scale dependencies, exhibit spatial and temporal lags with active feedback effects, and characterize qualities and quantities that may be expressed only in qualitative terms, simple definition may not be possible. Likewise, the situation may exist where there is a lack of theory relating to the problem or where the factors involved are sufficiently unique that they provide no basis for comparison. Given these conditions, the variety inherent to a given definition may explain an array of linkages and a level of uncertainty that frustrates a clear articulation of causality. Consider, for example, the problems facing the assessment of potential sites for high-level nuclear waste repositories. To determine their safety, decision makers must evaluate climatic, geologic, hydrologic, and even potential future human activities in and around these sites over a time horizon of 10,000 to 1 million years. Since there is limited understanding of radionuclide transport in the biosphere and the events that may cause the release of these materials, assessment must contend with staggering levels of uncertainty. Making a decision when confronted with conditions such as those outlined above requires a process orientation that permits a degree of experimentation in the phases that describe decision making. In the following section the phases characterizing the decision process and the issues that drive environmental decision making are examined.

Fundamental Decision Theory

Decision making has been defined as a process by which a person, group, or organization identifies a choice or judgment to be made, gathers and evaluates information about alternatives, and selects from among those alternatives (Carroll and Johnson, 1990). As a process, decision making explains a stream of thoughts and behaviors that also includes an element of risk. In the abstract, decision making illustrates a dynamic search for information that is full of detours, enriched by feedback, fueled by fluctuating uncertainty, and frustrated by indistinct and conflicting concepts (Zeleny, 1982). While this process may be difficult to explain succinctly and universally, it does exhibit a structure and share a set of formalisms that contributes to a basic understanding of the steps involved. This structure, however, cannot be captured by a decision tree alone, a decision table, a single mathematical function, or any other mechanistic artifact (Zeleny, 1982). Rather, its structure is functional, capable of generating its own unique path toward a solution. Therefore, a decision unfolds through a process of learning, understanding, information processing, information assessing, and careful definition of the problem and the circumstances involved. Applying information technology to support this process requires tools that enhance learning and understanding and provide a means to process and assess information as well. Therefore, in order for us to understand decision making, the processes leading to this solution need to be emphasized.

The decision-making process has been characterized in a number of ways (Inbar, 1979; Simon and Newell, 1982; Taylor, 1984; Carroll and Johnson, 1990). Carroll and Johnson (1990) describe decision making as a process consisting of the following seven distinct phases (Figure 2.2):

1. **Recognition**—realizing that there is a decision to be made.
2. **Formulation**—exploring and classifying the decision situation and forming a basic understanding of the relevant objectives and values.
3. **Alternative Generation**—producing a set of choices.
4. **Information Search**—identifying the attributes or properties of the alternatives under consideration.
5. **Judgment and Choice**—evaluating and comparing alternatives.
6. **Action**—taking action based on the decision.
7. **Feedback**—receiving information about the outcome of the action that permits changes in substantive knowledge and decision rules.

Using a less detailed description, Zeleny (1982) explains decision making as a process of three stages: predecision, decision, and post-decision. According to

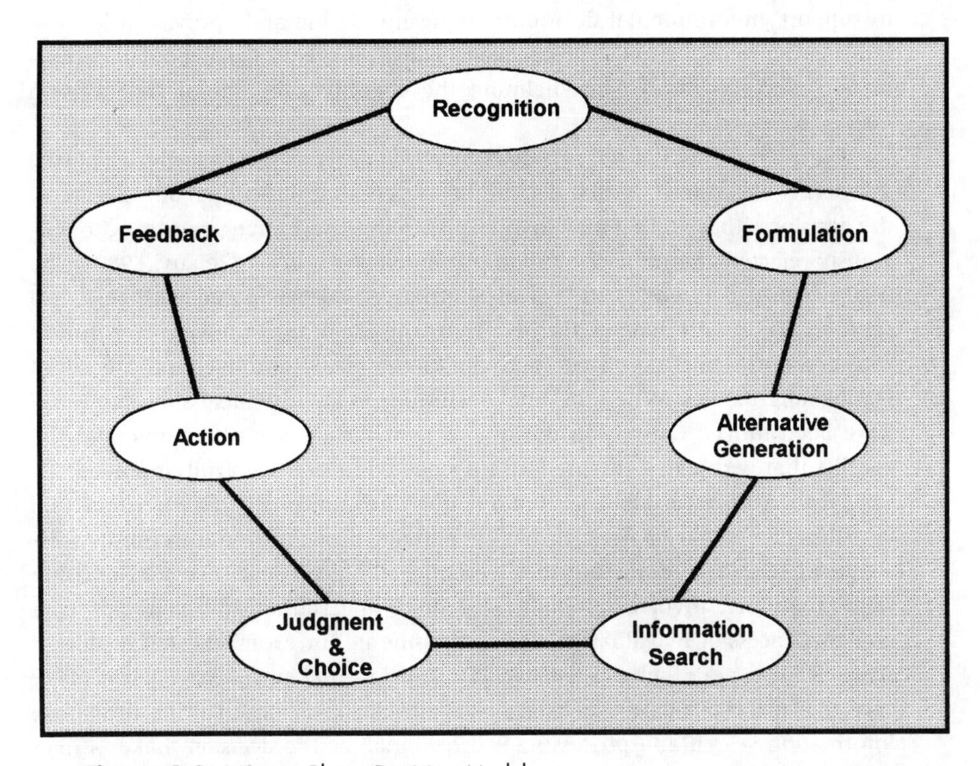

Figure 2.2 A Seven-Phase Decision Model

this typology, predecision emerges from a sense of conflict, which provides a decision-motivating tension, which encourages a search for an ideal solution. The source of predecision conflict is the general nonavailability of a suitable alternative and the infeasibility of the ideal alternative. Sensing conflict, the decision maker engages in a search for new alternatives that approximate the ideal. Each alternative is systematically evaluated by making choices or judgments, moving the search toward conflict resolution. The search for new information is a central activity in this stage. Initially during this phase information gathering and evaluation occur in a highly objective and impartial manner. However, as the decision maker realizes that additional information fails to reverse or appreciably influence the established order of preferences, the process becomes more subjective (Zeleny, 1982).

Once subjectivity enters into information gathering and evaluation, only select pieces of information are admitted. In fact, some information is consciously or unconsciously ignored while other information is reinterpreted or even dismissed (Zeleny, 1982). Experimental evidence supporting this behavior was provided by Festinger (1964) who demonstrated that the closer the alternatives are in their attractiveness and the more varied the acquired information is, the more information will be sought before a decision is made. Given the multi-faceted nature of the environment, an information technology designed to support environmental decision making must ultimately provide a level of information requisite to the problem and allow for the processing of large information stores without overwhelming the cognitive abilities of the decision maker.

During the decision stage of this three-step process, the concept of a partial decision is introduced. Partial decisioning describes a directional adjustment of the decision situation that may involve discarding alternatives, returning to previously rejected alternatives, or adding and deleting criteria (Zeleny, 1982). The final decision emerges as the set of alternatives are compared with the ideal solution. Those alternatives furthest from the ideal are removed from future consideration while those that approximate the ideal are moved into the set of feasible alternatives. When the final decision is made, the ideal alternative will have been moved entirely in the direction of the chosen alternative and the conflict that permeated the predecision stage will have been fully resolved.

Post-decision describes a situation not uncommon to the environmental decision maker. Once a decision has been made, post-decision uncertainty (dissonance) sets in to dominate the decision-making process. The post-decision stage, therefore, involves attempts to reduce uncertainty and characterizes a gradual process of reevaluation and reassessment. For example, after a route is selected for a new highway bypass, the decision maker/analyst tries to make sure that the route that was chosen was the "correct" one. During this stage, information search and processing is directional, as the decision maker strives to enhance the attractiveness of the preferred alternative and further reduce the attractiveness of those rejected. The objective search for information conducted during the predecision stage is now replaced by a selective confirmatory search that essentially supports the decision and verifies that the alternative selected was appropriate or correct.

A more concrete and stepwise model of decision making that combines the main procedural concepts discussed above was offered by Davis (1988). According to this interpretation, a decision is obtained by:

- formulating the possible alternatives and specifying the objectives,
- identifying the relevant influential factors,
- evaluating and analyzing each alternative,
- comparing and ranking the alternatives, and
- selecting the alternative that provides the best overall outcome relative to the objective.

This stepwise approach is illustrated in Figure 2.3. Decision making, however, is not simply a matter of phases, but is strongly influenced by the characteristics of the decision maker. These characteristics define certain styles of decision

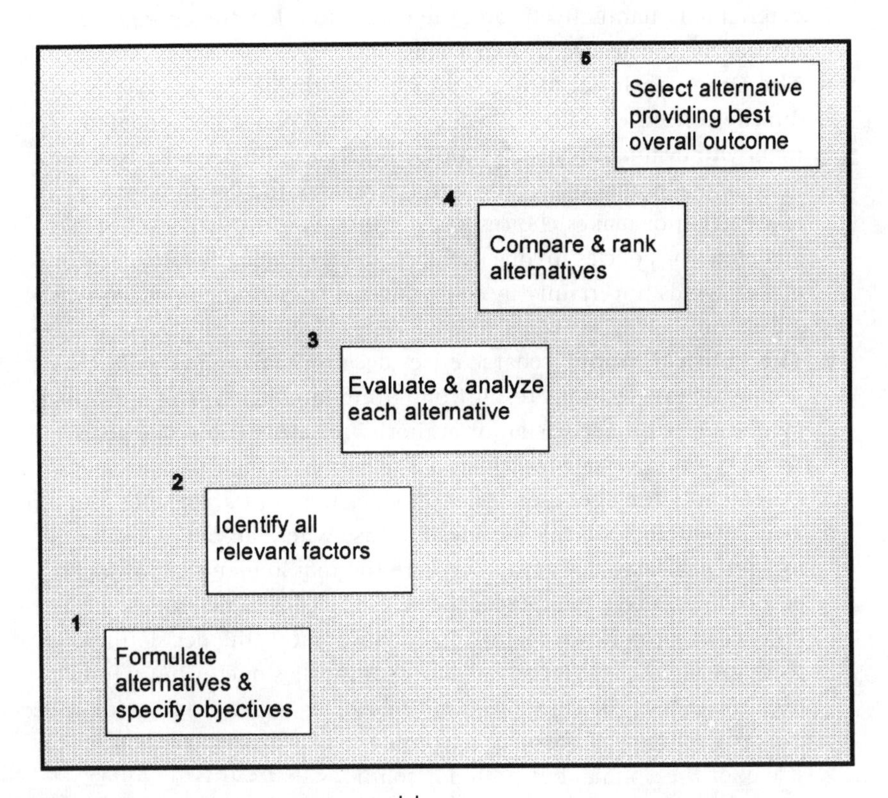

Figure 2.3 A Stepwise Decision Model

making that ultimately control how the decision process unfolds. Therefore, an information technology to support decision making must be flexible enough to accommodate these variations in how decisions are derived.

Decision Styles

The decision-making process described above, while instructive, fails to illuminate the reasoning methods individuals employ when approaching a problem. As noted by Davis (1988), each decision maker approaches a problem differently depending on factors such as background, experience, inherent psychological conditioning, and the situation surrounding the problem. Several classical models have been introduced to describe the philosophy or styles employed by decision makers when they approach a decision problem. Five of the more widely understood models that have relevance to the task of environmental decision making include (after Davis, 1988 and Bidogli, 1984):

- **The Rational Model**—views decision making as a structured process by which the individual or group systematically reduces the decision problem to a set of measurable quantities or qualities that influence the desired outcome. The comparative merits of each outcome can be determined quantitatively, and the goal is to select the alternative that defines the greatest measured value.
- **The Organizational Model**—implies that the decision maker is more directly concerned with following established policies or guidelines than with evaluating all relevant factors that influence a decision in a quantitative manner. According to this model, the decision maker takes action or makes choices based entirely on the active set of guidelines or policies involved. Subsequently, the individual decision maker avoids uncertainty by following a predetermined path to reach a decision.
- **The Political Model**—characterizes decision making in a political setting, where decisions result from group interaction. The individuals involved rely on persuasion or authority to satisfy subjective goals. There is no universally accepted best decision; rather, the ultimate goal is to identify the alternative that provides the most acceptable solution.
- **The Satisficing Model**—defines the case where an optimal solution to a problem may not exist, or where the requirements for a single best solution may be impractical due to the constraints imposed by time, costs, or personal factors. In such instances, the decision maker seeks an adequate alternative—one that satisfies one or more initial requirements of the solution—and then relies on feedback to improve the next solution, if possible. According to this model, decision makers strive for the optimal but settle for solutions that satisfy or suffice given certain constraints.

- **The Individual Model**—gives emphasis to the individual decision makers' idiosyncrasies. Each decision maker possesses a unique background, personality, and level of education, training, and experience, as well as other intangible qualities that define an individual's style. Combined, these traits direct the individual through the process of decision making, and in some instances preclude the individual from deriving a solution to a given decision problem.

Identifying the problem and specifying the factors involved are greatly influenced by the approach or model defining the "setting" in which the decision is made. From an information technology perspective, problems can be described as either programmed or nonprogrammed following the logic outlined by Simon (1960). Programmed decisions define problems that are repetitive or routine. Nonprogrammed decisions are unstructured and unique and require the exercise of judgment, intelligence, and adaptive problem-solving behavior. In the following section, the structure of problems confronting decision makers is examined, building upon these dual themes.

Characterizing the Problem Arena

Ultimately, decision making takes place within the context of a problem. Recognizing the problem and gaining insight into its features is central to any decision-making effort (Inbar, 1979). A decision problem can fall into one of three general categories (Gorry and Scott-Morton, 1971):

- structured,
- semistructured, or
- unstructured.

Structured problems tend to be well defined and clearly understood. Typically, they explain programmable tasks, which are routine in nature and can be addressed following a set of standard operating procedures. A simple example of a structured problem in an urban planning context is a building permit review. As a plan is submitted, it is compared against the codes for that structure in the zoning district where it will be constructed. If the standards set forth in the building and zoning regulations are met, the permit is issued. Structured problems exhibit little, if any, uncertainty, since the factors involved display no significant variability. Structured problems permit structured decision making that seldom requires expertise for implementation.

Semistructured problems are not as well defined as structured ones, and generally cannot be addressed adequately using standard operating procedures. While there may be some structured aspects to a semistructured problem, the factors involved display greater variability, thus increasing the level of uncertainty when a decision is made. Such is the case with a facilities location problem. Although the factors that direct a given location problem may be well understood, they can exhibit enough variability geographically as well as with respect to relative attractiveness to make it impossible to derive a clear and simple solution.

An unstructured problem is one that is unique in nature. It can either be nonrecurring or display characteristics that are highly variable. Because of their uniqueness, unstructured problems cannot be addressed using standard operating procedures. Instead, decision makers operating in an unstructured problem environment rely heavily on intuition, judgment, knowledge, and adaptive problem-solving behavior to compensate for the extreme levels of uncertainty that permeate all aspects of the situation. Siting decisions involving hazardous materials, issues related to the health effects of hazardous substances, and questions concerning long-term environmental impact and risk are examples of unstructured problems.

Uncertainty is a feature of even the most basic decision problem, and its influence in decision making and problem solving has not gone unnoticed (Lemons, 1995). Several of the more salient aspects of uncertainty central to decision making are examined in the following section.

The Role of Uncertainty

Human problem solving and decision making are most often performed in the context of uncertainty (Klein and Methlie, 1990). While the presence of uncertainty is widely acknowledged, its impact on decision making remains poorly understood and difficult to express (Morgan and Henruion, 1990). Recently, a taxonomy of uncertainty as seen from an environmental perspective was introduced (Suter et al., 1987). According to this conceptual design, uncertainty is divided into two principal categories: defined and undefined. Defined uncertainty explains the uncertainty intrinsic to the problem under consideration, whereas undefined uncertainty describes the inherently unknowable, which cannot be explicitly incorporated into the decision-making process. Although comparatively little can be done to reduce undefined uncertainty, defined uncertainty can be dissected further to reveal classes of uncertainty referred to as identity and analytical uncertainty (Suter et al., 1987).

By definition, identity uncertainty describes the uncertainty surrounding the identity of features and/or individuals impacted (affected) by a decision at some future point in time. Analytical uncertainty explains the uncertainty surrounding the prediction of environmental effects (impacts) emanating from the decision process. Within the context of prediction, three sources of uncertainty can be noted (Suter et al., 1987; Rowe, 1977):

1. uncertainty resulting from the approach used to conceptualize the problem and its context, which introduces model error,
2. uncertainty that manifests from the stochasticity intrinsic to human and physical processes, and

3. uncertainty associated with the problem of measurement, which contributes to parameter error.

Model Error

Environmental decision making frequently involves the application of either mathematical or statistical models to represent or explain process. Although the treatment of modeling as a decision-making tool will be examined in detail in Chapter 4, the question of model error relates to the general problem of correspondence between the model and reality. Because a model is simply a representation of a real feature or process, it will not capture all the details of the real phenomena. Therefore, irrespective of complexity, a model will exhibit some lack of fit between its representation and the relationship after which it is patterned. This lack of fit introduces several major sources of uncertainty, including:

- using too few variables to represent a large number of complex phenomena,
- selecting the incorrect functional form to express interactions among variables, and
- setting inappropriate boundaries for the components of the world that are to be incorporated into the model.

Stochasticity

Environmental decision making is unique in the sense that its primary focus is dynamic and subject to spatial and temporal variability. Regardless of scale, there exists a level of unpredictability or randomness that restricts the precision of measurement and ultimately the pattern of causality in any environmental system. This quality underlying the behavior of an environmental system implies that process is governed by probabilistic laws, where probability serves as a measure of human ignorance relative to the actual situation and its implications. Thus, in a system defined by "stochastic" relations, probability provides a means of describing the internal relationships of its components when its behavior is too complicated to be thoroughly comprehended.

Measurement

The environment, as defined earlier in this chapter, is characterized by conditions, quantities, and qualities that are difficult, if not impossible, to measure precisely. Often environmental variables can be expressed only in very qualitative terms that introduce a level of inexactness, imprecision, and vagueness that may defy simple quantification. In other instances, direct measurement of a variable may not be possible, requiring the decision maker either to employ estimates or to rely on surrogate variables to replace those lacking a metric.

The preceding treatment of decision theory and its defining concepts provides a rudimentary understanding of the process involved in making decisions. Environmental decision making, while conforming to the basic structure of the decision process as outlined, departs in several key ways from the general question of making a decision. In the next section, those aspects of the decision problem unique to the environment are examined.

Environmental Decisions

When considering the relationship between human population and the environment, the dominating focus of concern for well over a century has been the impact of human activities on the systems and subsystems that define the environment. Although this impact-centered perspective has produced a detailed literature documenting the environmental effects associated with human activities and contributed to a better understanding of human-induced environmental change, one critical feature of the environmental problem has largely gone unnoticed. This overlooked aspect of the environmental problem identifies the decision-making process that rests at the center of all human-caused environmental changes. Documenting studies describing the causal relationships that define modified or disrupted environmental processes, while essential for improving basic scientific understanding, are, therefore, insufficient by themselves unless the crucial knowledge they represent can be assembled into a form that can be applied in a decision-making context (Lein, 1993b). Because human-environmental impact is fundamentally the product of a human decision-making process, the documenting evidence of change that fills the literature is simply a by-product or consequence of the decisions that emanate from that process. This point is supported by Chechile (1991) who maintains not only that human decisions are at the core of most actions affecting the environment, but that society, in general, has made too many faulty decisions that fail to include an environmental component. To illustrate the need for environmental thinking in the decision-making process, Chechile poses several rudimentary environmental questions that society will be called upon to answer in the near future. A selection of these is listed below.

- What are the environmental impacts associated with a planned action?
- What are the long-term environmental risks associated with human technological progress?
- Does human action or inaction pose worrisome risks to health and safety?
- What psychological factors contribute to environmental exploitation and/or environmental conservation?
- What risks must be accepted by society and what is an equitable policy for distributing those risks?

- How can the aesthetic value of nature be balanced against the economic forces of development?
- How can society protect the environment while maintaining economic and technological progress?

Arriving at answers to these questions is not a trivial matter. Indeed, compared to other decision problems, environmental issues such as these introduced a level of complexity not found in other problem areas and foster divergent views on how environmental problems should be viewed and evaluated (Caldwell, 1987; Miller, 1993). In environmental decision making it is not uncommon for a comparatively straightforward issue to involve the physical, geographical, and biological sciences, economic, ethical, legal, and political factors, as well as technological, psychological, and social issues. Omitting any of these factors may render the decision process incomplete and will likely produce an unrealistic and ineffective solution that will fail at some level when implemented.

Effective environmental decision making, therefore, calls for a synthesis of knowledge, information, and experience from a cross section of subject areas. Developing integrative approaches to decision making requires a platform that provides a common basis for the exchange of ideas and information. Information technology can serve as this integrative platform, where the expertise of individuals from various disciplines can merge with data, analytical tools, and problem-specific knowledge to work toward solutions to problems. Without this type of integration, multidisciplinary efforts directed toward environmental problem solving remain a challenge for several reasons, including:

- an unwillingness to examine a problem from multiple perspectives,
- the methodological complexity of structuring a problem to illuminate its multidisciplinary nature,
- problems related to personal obstacles and disciplinary biases, and
- perceived lack of an integrative analytic platform.

These issues also point to the fact that environmental decision makers are inevitably guided by a set of assumptions and values that influence the outcome of their decisions (Caldwell, 1987). Value conflict is a common consequence, and the role of analytical science as the sole means of addressing environmental problems is increasingly unclear as holistic approaches based on intuition, judgment, and imagination are explored (Miller, 1993). However, broadening the conceptualization of a problem, rather than defining it more narrowly, demands consideration of spatial and temporal dimensions and a means of accessing a wider and more dynamic store of information to facilitate "holistic" thinking.

Introducing Time and Space

An environmental decision sets into motion a chain of causal events and actions. Environmental decision making is, therefore, not simply the issues surrounding the problem of making a selection among alternatives, but the more daunting responsibility of projecting the outcome of that selection onto the spatio-temporal situation surrounding the decision and its environment. To illustrate this point, consider the simple problem of purchasing an automobile. For the consumer, the selection of which car to buy may involve factors such as color, style, headroom, and the various options packaged with a given model. Cost factors then enter in as the consumer weighs the desired model against the loan payments, insurance rates, and down payment required. Before long, the decision maker focuses on which car offers the majority of desired features at a cost that can be afforded over the five years of the loan schedule given present and expected future earnings. Time directs the decision by forcing the consumer to extrapolate earnings as a means of evaluating the consequences of a given alternative. In the case of environmental decision making, time is often a more critical and central factor in the selection process, although it is a concept that can be difficult to articulate or quantify.

The environment is dynamic; change is an inherent feature of both the natural and human landscapes. Because the environment is always responding and adjusting to natural and human-induced disturbances—for example, fluctuations in climate, population numbers, nutrient levels, or land use and land cover arrangements—any decision that involves the environment contributes to that pattern of change and adjustment. Therefore, the effectiveness, quality, and correctness of an environmental decision has to be considered relative to how it will influence the pattern of environmental change over time. For example, a decision to rezone undeveloped land to a residential category will set into motion a sequence of alterations that may or may not have been anticipated when the original decision was made. Designating the area as residential will invariably lead to the decision to grant building permits. The construction of dwelling units will trigger an increase in population and vehicular traffic, generate an increased demand for water and energy resources, produce solid waste and sewage, and alter land values and tax rates in adjacent areas. These effects are cumulative and may not be realized immediately, but you can certainly see how the "simple" decision to rezone land sets into motion a cycle of cumulative causation that stands to affect not only the characteristics of the site, but potentially the surrounding environment as well, over an indefinite time horizon.

From this simple example, it can be seen that the process of change stemming from an environmental decision is difficult to characterize accurately because some effects associated with a given human action may be discrete while others may be continuous with respect to time; and when viewed geographically, some effects may be spatially confined while others may "leap frog" or be diffused over a wider region. Making an environmental decision, therefore, is not unlike starting a snowball rolling down a hill. As the decision and its ramifications continue to unfold, the issues and possible effects have the potential to increase in magnitude and importance, just as a snowball picks up speed and snow as it rolls along. This snowball analogy demonstrates the time-space dilemma that has become a critical issue in environmental decision making and exemplifies the fact that a decision made today can set into motion a chain of events and consequences that are realized in a heavily time-dependent manner.

Factoring the consequences of a decision into the process of making a selection among alternatives extends the decision problem well beyond the descriptive models that define decision making. However, adopting an impact-centered approach to decision making is very important, particularly when considering the issues that influence environmental quality. From an environmental perspective, the impact of a decision on environmental quality can be viewed using the same concepts commonly applied in risk management and can introduce the notion that a decision can display effects characterized as:

- acute,
- chronic, or
- synergistic/cumulative.

An acute effect is one that is recognized immediately following the implementation of a decision. Such effects might range from the immediate loss of habitat following the decision to clear woodland to the displacement of a population following the route selection for a new highway. A chronic effect is one that does not materialize until a sufficient period of time has lapsed following the decision. The key to a chronic effect is the length of time the environmental system is exposed to the changed conditions or to the conditions that promote change. For example, the decision to site a new reservoir to meet the demands of an urban-based population will promote microclimatic and hydrologic changes near the site that may, after a period of years, facilitate the introduction of exotic plant species into the region or disrupt the nesting patterns of indigenous birds. Synergistic or cumulative effects recognize the potential for a decision to introduce changes that appear insignificant when considered in isolation, but that produce an outcome that was unanticipated or unknowable when combined with other events or decisions. For instance, a city may decide to restrict growth. By doing so, growth effects may be transferred to surrounding communities, forcing civic leaders to draft policies to respond to the increased developmental pressures. With restricted growth, land market forces inflate home prices and rents as communities in the region begin to experience increased homelessness. Additional approaches to assist the decision makers conceptualize time introduce the concepts of direct, indirect, and time-delayed effects as governors of the selection process. With these considerations a wider range of cause and effect is introduced that requires data and a means to examine process that facilitates a better understanding of causality as a function of a given decision problem.

Together with time, environmental decisions share an inherently geographic dimension that is frequently undervalued in the decision process. Yet any environmental decision ultimately changes the spatial arrangement of that environment and the distribution of features that comprise it. In very simple terms, when an environmental decision is realized somewhere on the planet, it assumes a spatial expression. The spatial expression defines a unique geographic location, displays scale, and interacts with other elements of the landscape it occupies. It also assumes a form (morphology) and becomes a visible feature of the landscape in some way.

The realignment of environmental decision making to include the spatial dynamics and consequences surrounding a given decision problem has not received widespread attention. Environmental impact assessment is perhaps the best mechanism presently available for disclosing the spatial and temporal aspects of an action to decision makers, although neither is a required element in a documenting study.

Spatial thinking requires placing the decision problem and its alternatives in their geographic context, or setting. To the decision maker, this implies the innate ability to visualize the environment into which the decision will be implemented. Here, the problem and its features are "superimposed" on the landscape and the "fitness" of a given alternative is examined relative to the characteristics of the site and the situational factors that explain its regional setting. At issue is the well observed fact that environmental effects migrate and diffuse spatially, introducing impacts well beyond the locale originally targeted by a decision. This aspect of the decision problem introduces the concept of spatial scale as a complicating factor in the decision-making process. Although spatial scale can been expressed in a variety of ways, the most basic definition categorizes scale relative to the concept of resolution and explains scale in much the same way that a lens focuses on its subject (see Figure 2.4). Using this characterization, phenomena can be defined along a continuum that moves from the general to the specific using the familiar terminology of:

Macroscale
Mesoscale
Microscale

At each of these levels certain spatial phenomena and processes become visible as the mosaic of the landscape reveals more detail. This conceptualization of scale requires not only a widening of the geographic scope of the decision maker, but also knowledge and information regarding environmental conditions at varying levels of generalization together with a means to deliver that information (Wessman, 1992). Recognizing scale as an elastic concept demands a clearer definition of scale influences and a means to synthesize multi-scaled environmental systems into an organizing analytic framework (Lein, 1995; Wessman, 1992; Turner et al., 1990).

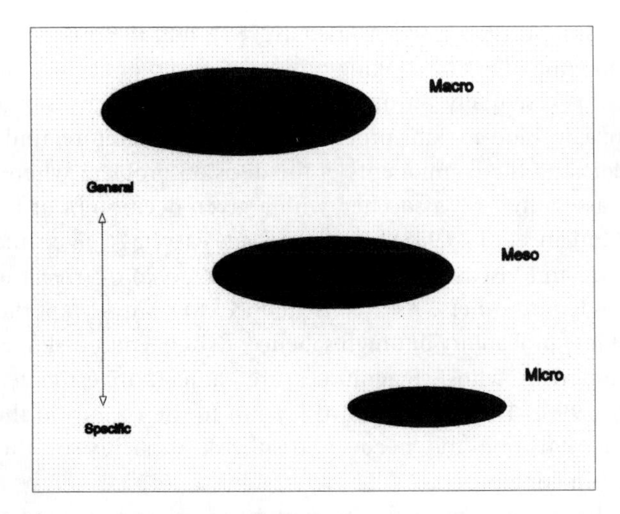

Figure 2.4 Spatial Scale and Spatial Resolution

One approach to the problem of scaling complex environmental systems involves the application of Hierarchy Theory. As a means of dealing with complex, multi-scaled systems, Hierarchy Theory dissects a phenomenon out of its complex spatiotemporal context and focuses inquiry on that phenomenon at a single time-space scale. This enables the problem to be defined more clearly and enhances selection of the proper "system" to emphasize in an analysis. Detailed treatments of Hierarchy Theory and its application to environmental problem solving can be found in O'Neill (1988) and O'Neill, DeAngelis, Waide, and Allen (1986). However, regardless of conceptualization, environmental decisions cannot ignore the presence of risk and its effect on the decision process.

The Question of Risk

Any decision contains an element of risk, since there is always the chance that a given selection, no matter how carefully derived, will fail to achieve the desired result. With respect to environmental decision making, there is also the possibility that the decision made to solve one problem will create several others over time or across space. Risk, therefore, is deeply embedded in the environmental decision-making process, and although it cannot be eliminated, it can be understood and minimized to some degree.

Over the past two decades a detailed literature covering the topic of risk assessment and its role in environmental management has developed. This literature explores the concept of risk from a variety of topical areas including ecological risk assessment (Suter et al., 1993; Bartell et al., 1992), environmental risk management (Lein, 1992; Whyte and Burton, 1980), risk modeling (Hunsaker et al., 1990; Fiksel, 1990; Cohen, 1984; Lehr, 1990), natural hazard risk assessment (Petak and Atkisson, 1982), and the link between risk and the environmental decision-making process (Covello et al., 1987).

Risk is essentially an outgrowth of the uncertainty surrounding the outcome of a decision-making process. Its presence, therefore, underscores the need to understand the components of the decision problem where risk is prevalent and to assess the risk associated with a given decision prior to its implementation (Morgan et al., 1985). This is not an easy task given the observation by Zeckhaser and Viscuss (1990) that society tends to overreact to some risks and virtually ignore others. As a consequence, too much importance is placed on risks of low probability but high salience, risks of commission rather than omission, and risks whose magnitudes are difficult to estimate (Zeckhaser and Viscuss, 1990). These aspects of risk point to the confusion the concept generates and illustrate the need to treat risk explicitly in the environmental decision-making process.

In the majority of cases, the concept of risk is explained as the probability (likelihood) that an adverse effect will occur. The adverse effect is typically characterized as a hazard that can be described as either a natural event or an event of anthropogenic origin. Regardless of description, environmental risks share several important features (in addition to their probabilistic nature) that can be used to categorize environmental risks as a set of related phenomena. These features have been summarized by Whyte and Burton (1980) to remind the environmental decision maker of the importance of factoring risk into the decision process. These essential considerations include the notion that risks:

1. involve a complex series of cause and effect relationships that tend to be connected from source to impact by pathways that include environmental, technological, and social variables that need to be modeled and understood in context.
2. are connected to each other, suggesting that several risks occur simultaneously within the same geographic area, which demands the ability to compare them and make trade-offs or balancing decisions regarding how much of one risk to accept in relation to another.
3. are connected to social benefits such that a reduction in one risk usually means a decline in the social benefits to be derived from accepting the risk.
4. are not always easy to identify, implying that identification frequently occurs long after serious adverse consequences have been experienced.
5. can never be measured precisely owing to their probabilistic nature; therefore, quantification is always a question of estimation to some degree.
6. are evaluated differently in social terms, meaning that a risk considered serious in one location may be considered unimportant in another,

underscoring the need to understand why similar procedures of risk evaluation give rise to dissimilar conclusions.

For the most part, reasoning about risk and applying estimates of risk in the decision-making process have developed following two contrasting approaches. The most common approach centers around the normative model of scientific reasoning based on probability theory, while a newer and less widely applied method employs techniques of human reasoning that draw on symbolic logic and symbol manipulation (Lein, 1992b). According to the tenets of scientific reasoning, risk forms from the probability of a consequence (A) given that an event (B) has occurred. This relationship between consequence and event can be expressed as a conditional probability where

$$p(A|B) = p(A \sim B)/p(B) \tag{2.1}$$

or, in the case of logic tree analysis, as a Bayesian probability drawn from a sequence of events where

$$p(A|B) = p(A) * p(A|B)/\sum(B_n * p(A|B_n)). \tag{2.2}$$

In other instances, mathematical expectation may be employed to predict the likelihood of an event (E) from a set of variables (X), such that

$$E = a_1X_1 + a_jX_j + \ldots + a_nX_n. \tag{2.3}$$

Human reasoning concerning risk and its estimation, however, follows a less rigorous model based on the supposition that people (decision makers) tend not to think in terms of probabilities in the classic sense. Rather, decision making involves the use of imprecise, inexact, or vague concepts that take meaning within the context of a spoken language. Consequently, an underlying ambiguity permeates all aspects of any decision-making process and can be attributed to the nature of language and how natural language concepts are applied. Here, reasoning about risk relies heavily on the knowledge, behavior, and prior experience of the decision maker to manage ambiguity and derive a meaningful solution. Indeed, the theory of human problem solving illustrates the fact that decision makers in general apply very few formal principles, violate normative rules, and rely on heuristics tailored to the semantic context of the problem (Klein and Methlie, 1990). Therefore, an incongruous situation materializes in risk assessment when normative models of probability cannot be estimated or when risk estimates based on normative models must be interpreted by a decision maker who reasons in terms of possibilities rather than probabilities (Apostolakis, 1990).

In addition, the complex issues surrounding the multidimensional nature of the environment suggest that it is often necessary to distinguish between empirically based findings of risk and those that are the product of human judgment. This is particularly the case in situations where empirical evidence is lacking—that is, where risk estimation relies on human judgment and the use of subjective probabilities (Fleming, 1991; Bonano et al., 1989; Wright and Ayton, 1994). In these instances the absence of a formal method of integrating human judgment into the risk assessment process simply adds to the uncertainty surrounding the original decision problem. Without such a mechanism, a clear understanding of the uncertainty attributable to human judgment cannot be achieved.

The presence of uncertainty in risk-based decision making has encouraged development of three contrasting approaches to risk assessment with little allowance for articulation between their findings: (1) empirical risk assessment based on scientific evidence, (2) model-based assessment employing predictive models in place of empirical evidence, and (3) qualitative risk assessment drawing on heuristics, human judgment, and qualitative reasoning (Fikel, 1990). While each method can produce contrasting levels of precision, an element of uncertainty remains whether it develops out of the stochasticity of the processes involved, the parameters and variables selected to drive predictive models, or simply as a function of human ignorance.

Improving risk assessment and the general process of incorporating risk concepts into environmental decision making requires a bridge that can connect the probabilistic definitions of risk based on scientific reasoning to the qualitative interpretation of the concept drawn from human judgment and reasoning. With such a device, a more formal method of decision making with uncertain or inexact information can be realized. Facilitating reasoning under conditions of uncertainty, however, places new emphasis on the representation of uncertainty and the ability to combine and draw inferences from uncertain information. One approach to bridging the gap between probabilistic decision models and the inexactness introduced by human reasoning that has demonstrated its utility in the area of environmental analysis and decision making is based on the application of fuzzy logic and the theory of fuzzy sets.

Fuzzy Decision Making

Environmental decision making is similar to any other decision process in that it takes place in a setting where the goals, constraints, and consequences of possible actions are not known precisely (Bellman and Zadeh, 1970). Furthermore, the complex interactions of factors critical to the problem are so integral to the knowledge of the decision maker and so frequently defined only in subjective and qualitative terms that communicating their significance is a common source of frustration and confusion (Lein, 1990). However, because environmental decision making requires its practitioners to operate behind the backdrop of uncertainty, where relevant facts may be incomplete and motives or goals vaguely defined, communicating vagueness, uncertainty, and imprecision is essential. The qualitative and subjective aspects of the decision process, which define the "fuzziness" of the problem, may otherwise render the most carefully crafted decision model inadequate.

Traditionally, imprecision has been communicated quantitatively through the use of probability theory techniques and concepts. This form of expression, however, carries the implicit assumption that imprecision is the same as randomness. Although randomness is an important component of uncertainty, it has been suggested that more careful differentiation is needed between randomness and the type of imprecision or inexactness that manifests itself as "fuzziness" (Zadeh, 1965; Bellman and Zadeh, 1970; Zadeh, 1978; Gaines, 1984). Having made this distinction, decision makers can better understand the sources of imprecision and realize improved decision models.

For practical application, fuzziness has been defined as a type of imprecision or vagueness that originates in language and characterizes phenomena, events, concepts, and features that cannot be precisely defined or measured. An excellent treatment of fuzziness and its representation using fuzzy logic and fuzzy set theory with emphasis on spatial analysis and planning can be found in Leung (1988). In a natural language many concepts lack precise definitions or explain relations without sharp boundaries separating them. Common concepts such as near, far, hot, cold, short, and tall are examples of this unique feature of language. These concepts defy exact definition, and the transitions implied by a set of objects moving from short to tall, or hot to cold, have no clear divisions that organize features along the continuum these concepts explain. Concepts such as these represent qualities and quantities that are not precisely defined; they express approximations instead, which are fuzzy by definition. In the field of environmental decision making, concepts such as risk, suitability, and carrying capacity, among others, can be added to the list of fuzzy concepts. Fuzziness is a way of characterizing the inexactness of concepts or classes of objects that cannot be described in terms of sharply defined boundaries. Thus, unlike a simple two-valued logic system, where an object either is or is not a member of a defining class or condition, an object in a fuzzy set can display partial membership in the class of objects or conditions that defines it.

Reasoning with fuzzy concepts is accomplished through the application of fuzzy logic. As explained by Zadeh (1988), fuzzy logic is concerned with the formal principles of approximate reasoning. Fuzzy logic, unlike classical logic, attempts to model imprecise modes of reasoning and find an approximate answer to a question based on a store of knowledge that is inexact, incomplete, or unreliable (Zadeh, 1988). Although fuzzy logic was introduced more than three decades ago, its application in environmental management and decision making has not been widely embraced. Notable exceptions include demonstration studies in the areas of site selection (Horsak and Damico, 1985; Anandalingam and Westfall, 1988); land evaluation and suitability assessment (Chang and Burrough, 1987; Sui, 1992); risk assessment (Kangari and Boyer, 1992;

Lein, 1992); and environmental impact analysis (Wenger and Rong, 1987). Beyond the arena of environmental analysis, fuzzy logic and the theory of fuzzy sets have been among the most widely researched and debated topics in the decision sciences (Zimmerman et al., 1984; Bezdek, 1987; Klir and Folger, 1988; Zadeh and Kacprzyk, 1992).

The central theme of fuzzy logic is the assertion that a variable can describe partial membership in a class. Objects can therefore have graded participation in a set that is inherently fuzzy. These inexactly defined concepts, referred to as fuzzy sets, have the property

$$A = \left\{ \left[X,\ \mu_{a(x)} \right] : x = X \right\} \tag{2.4}$$

where $\mu_{a(x)}$ expresses the degree or level of membership of object (x) in class (A). Fuzzy sets are therefore defined by means of a membership function (μ_a), which assigns each element (x), describing a given quality, condition, or value, its degree of compatibility or participation in the concept represented by (A). The level of membership an object (x) may assume in a fuzzy set can range from 0.0 to 1.0. According to the principles of fuzzy logic, an object whose membership is 0.0 defines the condition of perfect nonmembership in the class. A value of 1.0 defines perfect membership in the class of interest. The interval between 0.0 and 1.0 characterizes the possible values an element in (x) can explain relative to its degree of membership in the defining class (A). In general, the form of the membership function for a fuzzy variable can be determined by the relation

$$\mu_{a(x)} = \begin{cases} 1 & : \text{ if } x \geq \text{an upper threshold} \\ (x - \text{low})/\text{range} & : \text{ if low } < x > \text{upper} \\ 0 & : \text{ if } x \leq \text{a lower threshold} \end{cases} \tag{2.5}$$

Equation 2.5 states that membership in set (A) for the variable (x) will equal 1.00 (perfect membership) if the value of (x) equals or exceeds some upper threshold or cutoff value, 0.00 (perfect nonmembership) if the value of (x) is less than or equal to a lower threshold or cutoff value, or an intermediate range (value) defining its level of participation in the set if (x) falls somewhere along the interval bounded by the upper and lower thresholds. Through the use of Equation 2.5, the extent to which a specific value of a variable is compatible with its illustrating concept can be expressed. For instance, the measured value of air or water temperature can be expressed in relation to its degree of membership in the concept "hot" or "cold."

Although critics of fuzzy set theory contend that this practice is similar to the task of assigning subjective probabilities, a fuzzy set should not be interpreted as a probability distribution. Instead, a fuzzy set, by definition, characterizes a possibility distribution where the variable expressed as a fuzzy member in a fuzzy set is defined relative to its degree of compatibility with its defining concept (Zadeh, 1978; Mantaras, 1990). Possibility theory, therefore, differs from classical probability theory because it employs two expressions to represent uncertainty: a degree of possibility symbolized as alpha(α), and a degree of necessity given as $N(p)$. Therefore, a proposition (p) is assigned a degree of possibility and a degree of necessity such that

$$N(p) = 1 - \alpha(p) \tag{2.6}$$

where $N(p)$ in Equation 2.6 can be interpreted as the level to which (p) can be considered as necessarily true.

A comparatively straightforward example of reasoning with fuzzy concepts in an environmental application was offered by Lein (1992a). In this study, focus was placed on the concept "safe" as applied to the representation of distance surrounding a potential hazardous waste facility. The decision question posed for evaluation was thought to reflect a typical problem faced by environmental planners and decision makers and highlights the fuzzy nature of environmental decision making. The question evaluated by this study was presented as:

> How far from a population center should a hazardous waste facility be located in order to be considered a "safe" distance away?

In (n) trials, the grade of membership of distance (x) being considered safe was obtained as the frequency of distances being scored safe by a sample of 131 respondents to a questionnaire. A membership function of "safe" distance was derived from the sample according to the relation

$$\mu_{a(x)} = \left(\text{number of positive responses to "}x\text{ is safe"}\right)/n \tag{2.7}$$

The results of this test produced the membership function illustrated in Figure 2.5. The form of this function is open-end left and defines a break point value of 100 miles as a distance value completely compatible with the concept "safe," while values in between suggest differing degrees of "safeness" with respect to distance. The membership function produced by this study was applied in a geographic information system and was used to generate a fuzzy surface representing "safe distances" that was used in concert with other siting criteria to direct the facility location problem.

The integration of a fuzzy approach with classical decision theory was explored by Bellman and Zadeh (1970), Yager and Basson (1975), Kacprzyk (1982), and Zimmerman (1987). An excellent treatment reviewing the basic issues concerning the application of fuzzy set theory to decision analysis can be found in Spillman and Spillman (1987). When considering the adoption of a fuzzy approach to decision making, the decision problem is represented in mathematical terms and expressed as follows:

1. a set (A) of all alternatives,
2. a set (O) of all outcomes,

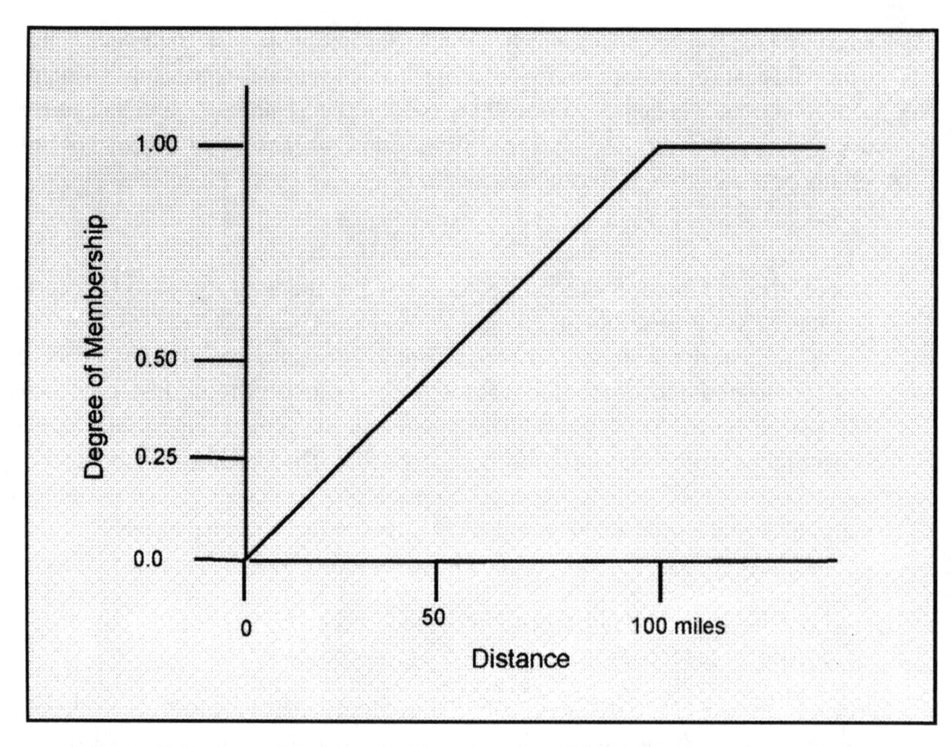

Figure 2.5 Generalized Membership Function of the Concept "Safe Distance"

3. a set (Z) of all possible states,
4. an outcome function,
5. a set of payoffs (G), or
6. a payoff function.

The method introduced by Bellman and Zadeh (1970) is based on the presence of a set of alternatives (A), a set of objectives (O), and a fuzzy set representation of the alternatives (O_i). Each alternative in the fuzzy set is defined in terms of a membership function that explains the degree to which each alternative meets the objective. This relation can be symbolized as (O_i). Given a decision criteria that the selected alternative must (should) satisfy, expressed here as

$$O_1 \text{ and } O_2 \text{ and} \ldots O_n, \tag{2.8}$$

by deriving a fuzzy set definition of "and" a decision function can be obtained that takes the general form

$$D = O_1 \cap O_2 \cap \ldots \cap O_n, \tag{2.9}$$

which can be represented using fuzzy operators as

$$D = \min\left(O_{1(ai)}, O_{2(ai)}, \ldots, O_{n(ai)}\right) \tag{2.10}$$

where D explains a fuzzy subset of the set of alternatives (a). Determining the "best" decision according to this approach is accomplished by selecting the alternative with the highest membership in (D).

An alternative method was suggested by Yager (1977) and refined by Yager (1982). This technique begins by defining a set of objectives and alternative actions that must be reconciled. With these sets so defined, each alternative is assigned a degree of membership relative to the set of objectives. Then the decision maker conducts a pair-wise comparison of the objectives by rating them on a ratio scale. The rating produces a matrix $[N]$ that is defined as

$$N = \left[n_{ij}\right] \tag{2.11}$$

where $n_{ij} = \Psi_i$, and explains the number of times objective (i) is rated more important than objective (j).

The matrix $[N]$ is analyzed to determine the unit eigenvector corresponding to the maximum eigenvalue using the method introduced by Saaty (1978). The result provides normalized weights (w_{ij}) for the objectives that are then multiplied by n to produce the final objective weight. The preferred alternative is then selected as

$$\max\left[\min\left(\Lambda_{ij}^{wi}\right)\right]. \tag{2.12}$$

This technique was extended by Yager (1982) to include representation of the impact of the decision-making task on the decision maker. This extension introduced the use of tranquility and anxiety measures based on the use of alpha-level sets. According to Yager (1982), an alpha-level set is a conventional (non-fuzzy) subset derived from a fuzzy set by thresholding the value of the fuzzy membership function at the level α, where $\alpha(0, 1)$.

The formal definition of the hardened alpha-level set for a fuzzy set (S) with a membership function $\mu_{(a)}$ is provided by Spillman and Spillman (1987) as

$$\mu^{\alpha}{}_{a(x)} = \begin{cases} 1 & : \text{ if } \mu_{a(x)} \geq \alpha \\ 0 & : \text{ if otherwise} \end{cases} \tag{2.13}$$

Given the situation where the decision maker must select from a set of alternatives (A) where the decision function (D) is a fuzzy subset of (A) and $D(a)$ measures the degree to which alternative (a) satisfies the overall decision criteria, tranquility explains the emotional or psychological confidence surrounding selection of the best alternative. Thus, if one alternative stands out above all others, the correct selection can be made with great confidence. If, however, all alternatives satisfy the decision criteria to some degree, decision making can produce anxiety. Following this logic, tranquility is expressed as

$$T(D, A) = \int_0^{\max} \text{Card}(D_a)^{-1} \alpha \qquad (2.14)$$

where α_{max} is the maximal grade of membership of any alternative (a) in (D) and CARD (D_a) is the cardinality of the alpha-level set of D. Anxiety, which is defined as the inverse of tranquility, is measured by the relation

$$K(A, D) = 1 - T(A, D). \qquad (2.15)$$

Risk, uncertainty, imprecision, and inexactness are all functions of the information used to support decision making. Those aspects of information germane to the environmental decision-making process are discussed in the next section.

Information: The Vital Ingredient

Environmental decision making does not occur in a vacuum. Defining the problem, identifying alternatives, visualizing spatial association, considering risk, and quantifying uncertainty all require information. As noted earlier in this chapter, information search is an essential component of the decision-making process, and the task of translating raw data into meaningful information is perhaps the most critical aspect of that process and one that greatly influences the quality and effectiveness of an environmental decision. It is not surprising, therefore, that so much effort is expended in attempts to improve the availability of environmental information and the methods and techniques of environmental data analysis (Hutchinson, 1992). Common sense would suggest that more informed judgments have a greater chance of achieving desirable goals than do less informed judgments.

Several activities are central to this theme. Chief among them is the need to determine the information requirements of the decision maker as a function of the decision problem. Certain information needs are shared by a range of environmental decision problems. A generic list of common environmental information needs is presented in Table 2.1. These information needs can be applied to a definable set of decision tasks as suggested in Table 2.2. A long-standing problem facing the decision maker has been acquiring fingertip access to this information and recognizing that the type of data or information incorporated into the decision-making process is ultimately determined by the problem and the objectives that drive the decision process. Three general approaches to assist decision makers in determining their actual information needs have been introduced (McLaughlin and Coleman, 1990; Hutchinson, 1992). These can be summarized as:

1. **A needs-driven approach**—where user needs are determined through an assessment of information requirements, analysis, and reports produced for projects of a similar nature.
2. **An information mandate**—which identifies the type of information

Table 2.1 Common Environmental Information Needs

DATA CATEGORY	SAMPLE MAP INFORMATION
Areal Data	Demographic Areas
	Zoning Districts
	Tax Assessment Areas
	Emergency Service Areas
	Watershed Areas
Base Map Data	Control Points
	Elevation Benchmarks
	Elevation Contours
	Building Footprints
Environmental Data	Soil Coverage
	Bedrock Geology
	Land Cover/Land Use
	Floodplain Coverage
	Streams & Water Bodies
	Special Habitats
	Unique Features
Land Records Data	Lot Boundaries
	Land Parcel Boundaries
Infrastructure Data	Sewer Systems
	Water Systems
	Electrical Cabling
	Telecommunication Networks
Street Network Data	Road Centerlines
	Street Intersections
	Road Types & Capacities

that must be available if an agency or decision-making unit is to fulfill its mandate as defined by law, enabling legislation, or agency charter.

3. **A utilitarian focus**—based on an evaluation of the impact various types of information have on decision making through some type of cost/benefit analysis.

Lastly, Hutchinson (1992) offers a composite approach. Here, user needs assessment is conducted to identify the range of information that might be needed with the information types that are absolutely required. From this

Table 2.2 Typical Environmental Decision Tasks

APPLICATION AREA	DECISION PROBLEM
Area Mapping	Map display & analysis
Development Tracking	Analyze development trends & display trends
Emergency Response	Analyze frequency & location of emergency events
Facilities Management	Plan facilities expansion; support planning & maintenance operations; update, display, & assess facilities data
Facilities Location	Selection of sites
Land Development	Analyze land records data; process & update change
Planning	Display & assess environmental and land use data; support plan analysis
Permitting & Licensing	Process and track permits; data display and analysis
Transportation Planning	Assessment of road system; evaluation of optimal transportation routing

assessment, an evaluation of each information type is made that rank orders their comparative value or importance to the decision-making process.

Once the basic information needs have been determined, the utility and quality of that information must be considered. Several factors influence the usefulness and quality of environmental data and information. Paramount among these factors are considerations regarding:

Information Type—describing the format, form, and media used to carry information to the decision maker,

Availability—explaining the relative ease of access and location of information,

Timeliness—concerning the speed at which information can be delivered to the decision maker,

Accuracy—expressing the reliability of information and the level of error or uncertainty it contains, and

Cost—defining the "price" of information expressed in an economic sense and from a human resources, time investment perspective.

In addition to the above, the information needs of the environmental decision-making process also require innovative ways of introducing information to the decision maker. Maintaining and facilitating the flow of information and filling critical information gaps places a premium on the "information infrastructure" that has grown around and supports environmental management and decision-making efforts. As suggested by Hutchinson (1992), the basic premise of the argument for developing information technologies to assist environmental decision making is that investments in "information infrastructure" will produce a stream of benefits to an agency or jurisdiction responsible for managing the environment that will quickly exceed the costs of acquiring and managing that information.

Realizing these benefits is not automatic. Understanding the technology and effectively linking its capabilities to the appropriate task or application is the key to successful integration. Without this basic understanding the decision maker may never realize technology's full potential or implement technological innovations as they become available. This chapter reviewed the basic concepts of the decision problem to lay the groundwork for understanding. Chapter 3 builds on this groundwork by exploring the issues surrounding information management, the recent advances in information management systems, and their realization in geographic information systems connected to the decision-making process.

Geographic Information Systems

Computer technology represents a means of performing routine tasks quickly, precisely, and with unparalleled flexibility. This is particularly evident in geographic information systems (GIS). A GIS offers exciting opportunities to improve decision making, yet the information needed to do this is scattered throughout an ever expanding maze of papers, conference proceedings, books, and technical reports. The aim of this chapter is to pull that material together and provide a comprehensive treatment of GIS, focusing on the issues germane to the problem of environmental decision making. We begin by considering the relationship between data, decisions, and the need for information management.

Data, Decisions, and Information Management

Decision making is a data-driven process. Without useful and reliable data, effective decisions cannot be made. Therefore, it is not surprising that agencies, governments, corporations, and even private individuals expend tremendous time, effort, energy, and money "sensing" and acquiring data about their environments. Data collection is an ongoing activity, often describing a highly specialized and focused operation, and at other times explaining a more general, casual process. But regardless of which format or code it defines, data is essential to the task of assembling, directing, and evaluating alternatives and is a critical ingredient in the process of making decisions. However, data by itself is useless, unless it can be organized into something that communicates to the decision maker. This requirement implies that data must be transformed into meaningful information, yet this transformation process alone is insufficient. Data and information must also be accessible to those who need it, and this added requirement demands the capacity to store, manage, retrieve, and disseminate data as well. Taken together, these nontrivial aspects of data—transformation, storage, management, and dissemination—suggest the need for a system of data management, data manipulation, and data analysis equal to the complexity of the decision-making process.

The environment is a multifaceted, multidimensional, complex assemblage of human, physical, and ecological components. From a purely data-abstracted perspective, the environment can be conceptualized using the familiar "layer cake" model, and can be defined as a collection of variables (Figure 3.1). The need to acquire environmental data to guide decision making in this complex mosaic has contributed to the development of numerous data acquisition technologies ranging from traditional ground surveys to space-borne satellite sensors. These technologies facilitate a comprehensive depiction of the environment and have enabled decision makers to define a very rich "layer cake" relative to the available database from which they can draw.

Paralleling the developments in data acquisition technology have been the growing need for more and varied data and a heightened demand for improved methods of information management and analysis. With an ever-widening array of software and hardware products increasing in power, capability, and sophistication and decreasing in price, information technology has become an indispensable partner in the decision-making process. However, the emergence of information technology, while generally regarded as beneficial to the decision maker, has created two unique problems. First, the decision maker must understand the capabilities and limitations of this technology and, once understood, apply the technology appropriately. Secondly, once a technology is discovered

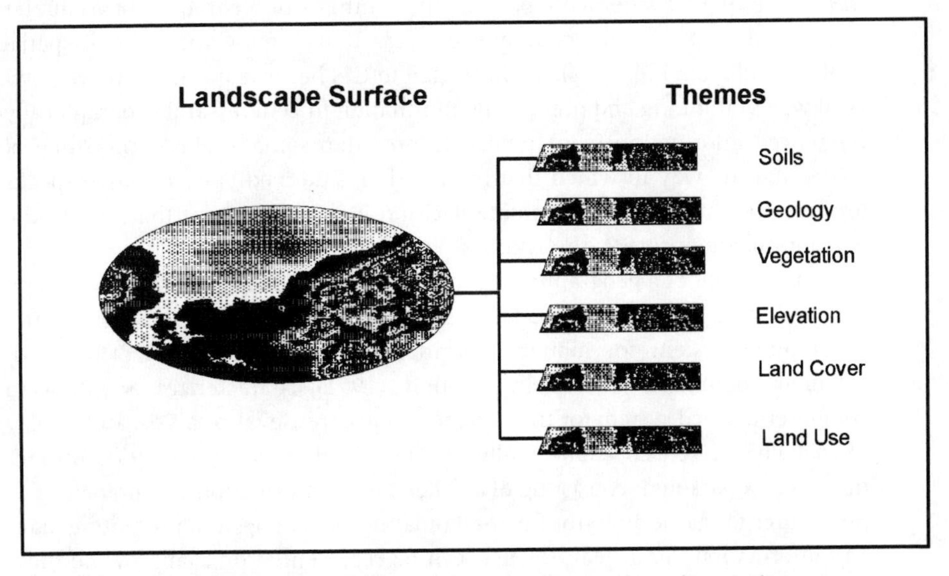

Figure 3.1 Layer Cake Model Describing Fundamental Landscape Themes

and understood, the decision maker must avoid the tendency to become absorbed in the technology at the expense of the problem that demands resolution.

An information systems approach to environmental decision making is one means of preserving the connection between the environment as explained by the layer cake analogy and the methods of data acquisition, data management, and data analysis that culminate in the eventual presentation of data as information. Of all the various information organizing schemes available, none displays the potential for meeting the requirements of the environmental decision maker or demonstrates its adaptability to a range of environmental problems more than the geographic information system (GIS).

GIS in Context

Since their emergence over three decades ago, computerized geographic information systems have received an extraordinary degree of attention, and they are still evolving. There are numerous examples of the development of GIS technology and its applications to environmental management, planning, and analysis. A cursory sample of this literature includes introductory texts (Starr and Estes, 1990; Aronoff, 1991; Burrough, 1988); environmentally focused works (Haines-Young, Green, and Cousins, 1993; Goodchild, Parks, and Steyaert, 1993); technical discussions (Laurini and Thompson, 1992; Huxhold and Levinsohn, 1995); applications reviews (Worrall, 1990; Masser and Blackmore, 1991); planning (Scholten and Stillwell, 1990; McCloy, 1995); and hazard and risk assessment (Carrara and Guzzetti, 1995; Stein, 1995). With such widespread interest in GIS technology, it is important to keep in mind that the essentials of GIS are not new. Rather, GIS explains the synthesis of techniques of computer graphics, data processing, and subject areas that employ methods of spatial analysis and mapping. People are attracted to GIS because it offers a more powerful way of thinking and presenting information in general, and geographically referenced information in particular. Information remains the cornerstone of a GIS, and the way in which that information is derived and presented speaks to the system's capacity to facilitate decision making. Without that link to the decision-making process, the GIS has no reason for being.

The concept of a geographic information system has been defined in several contrasting ways (Maguire, 1991). At its most fundamental level, a GIS defines a computer system for managing spatial data (Bonham-Carter, 1994). An expanded definition offered by Aronoff (1991) characterizes a GIS as a computer-assisted system for the capture, storage, retrieval, analysis, and display of spatially referenced data. Finally, Huxhold and Levinsohn (1995) consider the GIS as a paradigm consisting of a collection of information technologies and procedures for gathering, storing, manipulating, analyzing, and presenting maps and information about features that can be represented on maps. While these views of GIS differ in terms of specifics, they share one common theme, the requirement that GIS involves the integration of spatially referenced data in a problem-solving environment. Spatial reference is therefore a critical aspect of GIS that distinguishes it from other information systems and gives its users the ability to explore and analyze data with respect to a recognized system of location—that is, latitude/longitude, Universal Transverse Macator, State Plane, and so on.

In its capacity as a problem-solving tool, a GIS consists of four basic components that must function in a complementary manner in order to provide decision making with effective support. These components are illustrated in Figure 3.2 and include:

1. **The Physical Component**—defines the hardware and software elements that provide data and information handling and processing functions. The physical components of a GIS can be further divided into:
   (a) **The Hardware Environment**—explains the computer systems and related peripheral devices such as digitizers, printers, tape and disk systems, and plotters needed for the input, processing, and output of spatial information,
   (b) **The Software Environment**—defines the specific GIS package

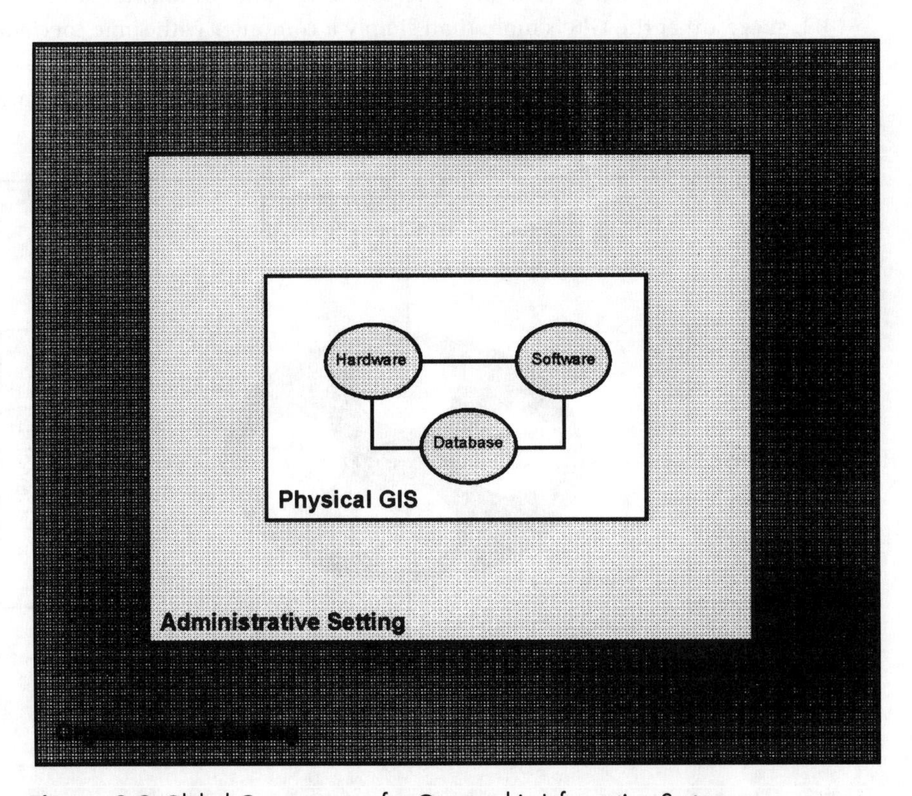

Figure 3.2 Global Components of a Geographic Information System

selected for implementation and related software systems required in order to realize design goals,

(c) **The Data Environment**—describes the logical structuring, management, and creation of data in the form of databases related to the needs of the decision maker.

2. **The Administrative Component**—explains the system of management, maintenance, and support required to oversee the operation of the physical aspects of the system. The administrative aspect of the GIS consists of several subcomponents including:
   (a) Database Administration
   (b) Application Specialists and Programmers
   (c) Database Security
3. **The Organizational Component**—describes the setting in which decision making occurs. This component provides the overall rationale for the system, directing its purpose, goals, and focus.
4. **The User Component**—characterizes the individual decision maker who must know what a GIS is, what it can do, and how to bring a specific problem to the GIS and obtain a satisfactory solution.

These four components are highly interconnected and, as illustrated in Figure 3.3, suggest that the GIS is more than simply a computer with some specialized software to run on it. Indeed, with respect to environmental decision making, the organizational setting and the users of the system may in the long run prove

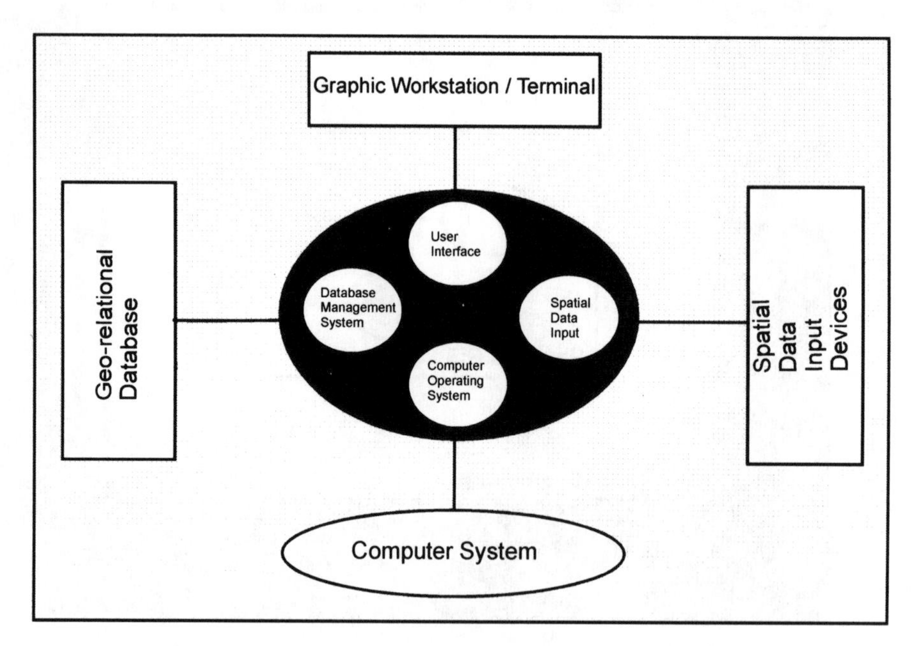

Figure 3.3 Physical Components of a Geographic Information System

to be more important than the technology simply because the organization and system users will ultimately determine whether the GIS is a benefit to the decision process or not.

In a critique of GIS, Aageenbrug (1991) explores this point and exposes the gap between the promise of GIS technology and its reality. In this review, Aageenbrug notes that GIS is not a panacea and calls for a more considered view regarding the applicability of GIS and its role in solving substantive problems. In a similar vein, Holmberg (1994) maintains that current GISs are employed primarily to reduce the costs and increase the quality of routine jobs. This observation suggests that current technology has been used at only a fraction of its potential. Extracting the full potential of GIS begins first by recognizing what GIS does well—its accepted benefits—and then moving on to explore pathways to expand its applicability.

The general benefits of GIS have been well stated in the literature and include considerations such as:

- Improved quality of information,
- Improved timeliness of information,
- Improved information flow,
- Improved efficiency,
- Improved productivity,
- Improved decision effectiveness, and
- Improved job performance.

Realizing the benefits of GIS, however, is not automatic. It begins by considering the factors that lead to the successful development and implementation of GIS technology. For purposes of environmental decision making, four critical ingredients demand careful attention: (1) system design and function, (2) data acquisition and database development, (3) analytic functionality and error, and (4) management, maintenance, and modification. Each of these ingredients is discussed briefly below.

System Design and Function

A successful GIS is created to fulfill one or more goals of an organization that recognizes the advantages of utilizing "geographic" data to support decision making. The concept of "geographic" data is central to this idea and implies that features on Earth's surface, whether they describe soil types, land uses, property lines, or mailboxes, are recorded with respect to their geographic location relative to a recognized coordinate system. Geographic features can be referenced to the surface, and the qualities or quantities that give them definition can be directly related to those features and represented in map form. Geographic reference gives data new meaning by relating a name, value, quality, or condition to an object that can be visualized and treated as a combined item (that is, a location that describes a certain attribute, or an attribute that resides at a given location). Thus, with GIS, spatial data can be organized, visualized, queried, combined, and analyzed in a manner not possible using manual techniques. None of this, however, can be realized without first establishing a purpose for the GIS and designing a system that will achieve the goals and objects that grow from that purpose (Huxhold and Levinsohn, 1995).

There are two basic paths to GIS design that reflect how the system and its purpose complement each other: focused and panoramic. A focused approach to design centers around a clearly specified application or purpose. An example of focused design might include a municipality that develops a GIS for property tax assessment. In this example, the requirements for efficient tax assessment as determined by the auditor's office establish the initial direction for system design. With development centering around a relatively high-priority problem, decisions concerning database design, functionality, and management are predetermined. In contrast to focused development is the panoramic approach. As the term implies, panoramic development has no single focus guiding design, but rather, adopts a wide view of GIS development that tries to encompass a range of application areas. The GIS, according to this view, is envisioned as a dynamic store of spatial data, adaptable to a broad range of applications, needs, and user requirements. In this sense, the GIS functions primarily as a repository of spatial data where users bring their problems to the GIS, which is equipped with generic functionality. With the panoramic approach, the system itself remains "open-ended." Thus, by capitalizing on this broad view of spatial data, options for future growth of the system are maintained and the system evolves in a modular fashion.

Regardless of whether development follows a focused or panoramic path, a commitment to developing the system must be sustained throughout the process. Assuming that a typical GIS will require a four-to-eight-year time horizon for development and implementation, maintaining the drive to construct and see the design process through to completion is a challenge (Cassettari, 1993; Huxhold, 1991). Constructing a GIS usually begins with a long-range problem to be solved and the intention to find all or a portion of its solution from the information generated by the system. From this point forward, the specifics of design take form. Physical design of the system involves several critical requirements, which must be articulated clearly, including:

- **Information Needs**—what the system is expected to deliver in the form of information products to its users.
- **Data Needs**—what users of the system require to perform their jobs and fulfill specific information needs.
- **Software Capacity and Functionality**—the range of software capabilities that are needed to deliver the required information, including specific analytic techniques, display options, and performance specifications.

- **Hardware Capacity**—the machine capabilities and associated peripheral equipment necessary for optimal software performance and information delivery.

Data Acquisition and Database Development

Data is the essential ingredient of any information system. Its accuracy, reliability, and accessibility are paramount to successful decision making. Although there is no single best approach to database development, the critical steps involved can be listed and their salient characteristics examined. The fundamental stages in GIS database development are outlined in Figure 3.4. As suggested by the outline, developing a GIS database centers around three primary activities:

- assessment of information and product needs,
- evaluation and collection of relevant data, and
- specification of a conceptual database.

Because the data in a GIS is varied and complex, understanding its physical representation in the database is an important link to using that data effec-

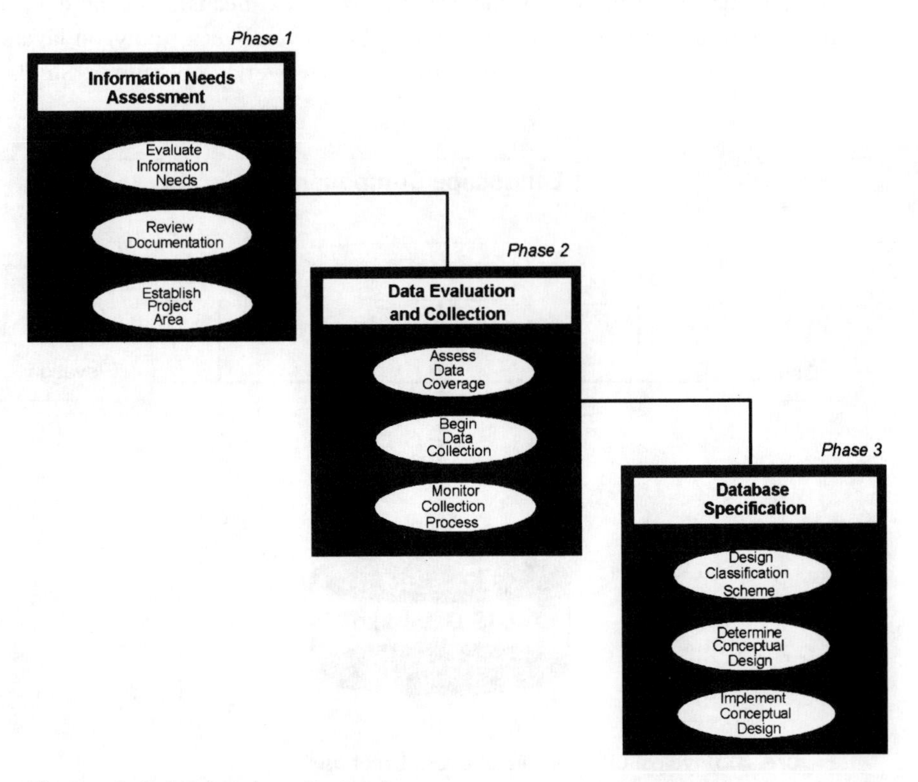

Figure 3.4 GIS Database Design Process

tively in a decision-making context. Very simply, the complexity of the real world must be abstracted and structured into a spatial data model that can serve as a formal means of representing its essential characteristics. This spatial data model is then translated into a selected data structure, which is then encoded into the appropriate file format (Peuquet, 1984; Goodchild, 1992). Presently, GIS technology recognizes two principal methods of representing spatial data: (1) raster format and (2) vector format. Although specific software systems handle geographic data differently, the concept of data organized into layers is a convenient way of visualizing spatial data, as it is structured into a database and returns to the layer cake analogy introduced to represent landscape complexity (Figure 3.5).

During the design process, the question of which mode of spatial data representation to employ is an important one when developing the spatial database. Burrough (1988) provides some useful insights to help answer this question. Vector methods have the advantage of a compact data structure that provides a good representation of real-world phenomena that can be topologically described with network linkages. This facilitates accurate graphics and makes retrieval, updating, and generalization of both the spatial and attribute data possible. The disadvantages associated with the vector model include the complex nature of the data structure. This complex means of spatial data representation introduces major limitations when several vector polygon layers or polygon and raster layers are combined through overlay operations. Similarly,

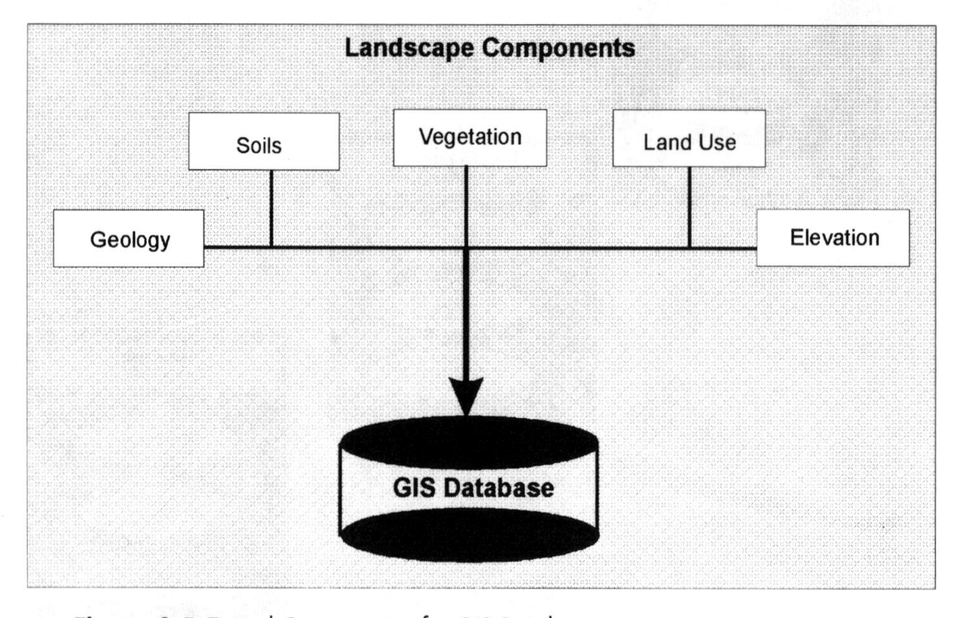

Figure 3.5 Typical Components of a GIS Database

analytic operations tend to be more difficult to perform since each feature of the real world has a different topological form. This renders some types of spatial analysis and filtering within polygons impossible.

By way of contrast, the raster method of data representation has the advantage of being a relatively simple data structure. This permits easy combination of data across layers and facilitates a wide range of spatial analytic techniques. The grid-cell design of the raster model also enhances simulation since all spatial units defined in the grid are the same size and shape. Raster methods, however, are problematic largely because they encourage the generation of large volumes of graphic data. Additionally, the use of the grid-cell representation of real-world phenomena risks serious loss of information through generalization, which can introduce scale and resolution conflicts that may fail to adequately represent complex features precisely. Consequently, raster systems generate products that are less cartographically accurate than those of their vector counterparts.

Once the representation issues have been resolved, the database begins to take shape. Spatial data organized to represent various themes selected from reality and stored either physically or conceptually as a series of thematic layers defines the database of the GIS. Navigating a database relies on a database management system not only to implement a specific data structuring scheme, but also to provide capabilities for data entry, editing, search, and storage. The relational model has become a widely employed data structure that is useful in handling attribute data in a GIS.

According to the relational data model, data is viewed as a two-dimensional tabular structure called a relation. Each relation is given a mane and represents a file that will hold data specific to a theme. As such, a relation is simply a two-dimensional table that contains a collection of data items that describe an object of the real world such as soil types, street addresses, or population characteristics (Figure 3.6). Items are connected to the relation through the assignment of primary, secondary, and cross-reference keys. The key provides a mechanism to organize data in the relation and facilitate search. Although the physical arrangement and ordering of data items in a relation can be arbitrary, keys provide an essential indexing system and a structure that permits efficient search and preserves logical connections among relations. Relations can be linked via keys using a relation algebra to enable more detailed and complex searches. Therefore, the naming of keys and their logical assignment are critical tasks during the database development phase. With a logical system of keys, quick and reliable data access and processing can be insured. Once these features are in place, the relational model can directly and implicitly represent a wide variety of associations among data items. This gives the relational data model heightened flexibility and makes this data structure comparatively easy to understand and implement.

Record Number	Street Name	Year Built	Building Type
100-01	15 Applegate Dr.	1990	DU
100-02	16 Applegate Dr.		VAC
100-03	17 Applegate Dr.	1987	DU
100-04	18 Applegate Dr.	1989	APT
100-05	19 Applegate Dr.		VAC
100-06	20 Applegate Dr.	1993	DU

Figure 3.6 An Illustrative Example of a Relational Data File

Analytic Functionality

With data assembled into the appropriate file structure to form the system's database, attention can be directed to the capabilities of the GIS and the analytic operations it can perform. When the question of functionality emerges, the GIS can be envisioned as a toolbox containing all the necessary routines required to process and manipulate spatial data to meet the information needs of the decision maker. The nature of this toolbox has been critically examined by Kliskey (1995) relative to the application of GIS as a planning tool in natural resource management.

Functionality in this context defines the capability of a system to perform functions involved in GIS data processing (Woodcock et al., 1990). Thus, when applying GIS to environmental decision making, it is essential to insure that the required analytic techniques are contained in the system or that the system supports an open architecture that permits development of new application programs. Considerations regarding functionality are the embodiment of the software component of the GIS and ultimately form the final expression of a systems value to the decision-making process.

Computer software, when installed in the machine environment, can be described in terms of levels, from the operating system at the lowest level, through special system support programs, to the upper-level application software (Antenucci et al., 1991). The software component of the GIS resides at this top level and typically consists of two integrated modules:

1. a "core" module of basic mapping and data management routines, and
2. an "application" module that performs specific mapping or geographic analysis routines.

Although individual software systems differ in terms of specifics, three operations are generic to this component:

- graphic processing functions,
- database management functions, and
- basic cartographic and geographic analysis utilities.

A general description of capabilities typically found under each of these functions is presented in Table 3.1.

Since the analysis of "geographic" patterns and associations is the driving force behind the application of GIS in environmental decision making, the ability of the system to perform complex geographic (spatial) analysis is essential, particularly if the potential of this technology is to be maximized. Although the majority of systems excel in the collection, organization, and visualization of spatial data, their ability to perform spatial analysis tends to be limited (Bonham-Carter, 1995). Recognizing the analytic limitations of the current generation of GIS, several authors have presented agendas for improving the integration of methods of spatial analysis with GIS (Rogerson and Fortheringham, 1994; Anselin and Getis, 1992; Goodchild et al., 1992; Openshaw, 1990). The drive propelling this research is in response to an increasing demand for systems that can do more. Presently, the analytic toolbox of a GIS consists primarily of overlay operations, the dilation (buffering) of spatial objects, neighborhood operations, tabular analysis, distance operations, and query and display functions (Burrough, 1986; Aronoff, 1991). A simple classification of detailing commonly encountered GIS functions is described in Aronoff (1991).

Management, Maintenance, and Modification

Effective decision making demands accurate and timely information. The database approach to decision making, capitalizing on the application of GIS, provides a centralized system for data sharing with little if any redundancy, and direct user access that is capable of meeting the diverse needs of the decision maker. However, once encoded into the database, data is static. The landscape that data represents is dynamic. Consequently, a level of management must be in place to initiate and oversee the maintenance and modification of the GIS to keep the system current and functional. Anticipating and planning for change is essential in this context. As GIS technology matures and the specific imple-

Table 3.1 GIS Functional Capabilities

GRAPHIC PROCESSING	DATABASE MANAGEMENT	MAPPING & ANALYSIS
Graphic Data Entry	Data Definition	Map Transformations
Interactive Digitizing		Coordinate Translation
Special Feature Entry	Data Entry	Projection Transformations
		Rubber Sheeting
Annotation Entry	Data Query & Search	
		Map Operations
Graphic Editing	Attribute to Map Link	Edge Matching
Feature Delete		Map Merging
Feature Modification		Windowing & Extractions
Line Generalization		
		Map Utilities
Graphic Display		Graphic Overlay
		Thematic Mapping
Graphic Plotting		Address Matching
		Overlay Analysis
		Map Analysis Utilities
		Buffer Generation
		Radius/Distance Search
		Measurement

mentation process ends, greater emphasis will be placed on operating and maintaining the system.

To assist in maintaining the long-term viability of a GIS, a management philosophy referred to as information resource management (IRM) has emerged (Huxhold and Levinsohn, 1995). This management philosophy provides an important connection between the GIS and the decision-making units that rely on it. According to Huxhold and Levinsohn, IRM is based on the principle that information is an asset requiring rigorous management. The underlying premise of this management approach to GIS is that the computing resources of an agency or organization, defining both hardware, software, data, and people, interact and undergo constant change. Through IRM, change can be managed in an orderly, cost-effective and appropriate manner by institutionalizing the use of computing technology on an organization-wide basis. Institutionalization, however, assumes that the appropriate data and technology infrastructure is available when called upon. Given this assumption, IRM encourages the integration of GIS concepts and information technology planning to direct and support the management process.

Once the GIS is operational, four critical factors influence and sustain the system's utility. These factors, listed below in order of relative importance, require a plan to insure that each follows a schedule or sequence timed to match and enhance the information requirements of the decision-making body.

- Database updating and error correction
- Software revisions and enhancements
- Hardware upgrading and replacement
- Application refinement and expansion

1. **Database Updating**—No element of the GIS is as important as the data comprising the database, and no element is as volatile. The accuracy, consistency, and completeness of data largely determine whether decision makers employ the GIS solution. Consequently, a detailed program must be put in place to plan and schedule systematic updating and revision of both the spatial and attribute data that form the GIS database. The specific schedule chosen is ultimately a function of the rate at which new data becomes available and the need for "current" data demanded by system users. Update schedules could follow and annual, quarterly, or shorter cycle depending on the type of data involved, the method of primary data collection, and related institutional factors.
2. **Software Revisions and Enhancements**—With each round of innovation in the technology of information, new developments in the software systems that pilot computers emerge at an ever-quickening pace. Improvements in the user interface, methods of data processing, functionality together with enhanced links to third-party software applications, and new machine architectures can create a "future shock" atmosphere with which management must contend. Basically, a strategy for adopting software revisions and critically evaluating software embellishments is needed in order to avoid adding capabilities to the system that are unnecessary, have limited utility to the primary goals of the system, or will be underpowered by the existing hardware environment.
3. **Hardware Upgrading and Replacement**—Like the software environment discussed above, computer hardware increases in sophistication by nearly one order of magnitude every 12 to 18 months. To keep place with technological progress, consideration must be given to the "appropriate" level of technology required to drive the system, maintain the overall goals and objectives of the GIS, and support decision making at an accepted level of performance.
4. **Application Refinement and Expansion**—Initially, the GIS may be

applied to a limited number of comparatively simple or routine tasks. As the system is learned and integrated into the decision-making process, users of the system develop a model of how the GIS works. At first this model explains a very general users' view of the system, describing each user's conceptual view of what is taking place within the physical GIS. As this model takes shape and deepens, it contributes to the refining and broadening of the methodologies imported to the GIS and encourages the development and testing of new problem-solving methods and applications.

Developing the GIS Solution

Although the role of GIS in environmental decision making has been widely accepted, comparatively little guidance is available detailing the specifics of applying GIS in a decision-making context. As Webster (1993) shows in a review examining GIS and its role in urban planning, there are four general categories that consider GIS and its connection to decision making in an indirect manner:

1. accounts of specific applications,
2. technical contributions advancing GIS technology,
3. organizational aspects of GIS, and
4. links between GIS and information science.

Although this literature suggests, in a general sense, how GIS contributes to decision making, the specific "how to" knowledge detailing the theory of GIS-based decision making is acquired primarily by means of illustration and proof-of-concept demonstrations. A more formal treatment of GIS-based decision making is therefore required.

Recalling the outline of decision-making theory in Chapter 2, the decision process for any given problem can be understood following a basic six-stage process involving: (1) problem identification, (2) goal setting, (3) evaluating alternatives, (4) generating a solution, (5) selecting a preferred solution, and (6) implementing and monitoring the decision. Connecting GIS to decision making, therefore, requires linking the capabilities of a GIS to a specific phase in this process. In this sense, the GIS provides critical "scientific" inputs to the decision process, which support an informational need of the decision maker. Webster (1993) provides an important discussion regarding the exact nature of this support. The thesis of Webster's argument is that each phase of the decision-making process can draw upon a specific functional capability housed in the GIS and that capability delivers information that moves the decision process along.

In general, Webster (1993) identifies three support functions provided by GIS that can be directly matched with a phase in the decision-making process. The functions identified should not be confused with the concept of functionality discussed earlier; rather, they explain data-to-information transformations that communicate specific realizations of a problem to the decision maker. The functions identified include: (1) description, (2) prescription, and (3) predication. As suggested in Figure 3.7, description provides data input during the problem identification, implementation, and monitoring phases of the decision-making process. Description in this context explains two central capabilities—measurement and documentation—that could include the production of mapped information or the presentation of information in tabular form. Through the application of GIS, critical attributes can be examined and their relative status can be evaluated. In addition, qualities or conditions can be measured and compared against a set of standards, and the patterns they reveal can be studied. As a complement to measurement activities, maps and other informational products can be generated to identify patterns in the data, illustrate an anomaly, and document what was discovered during an analysis. These capabilities enable the spatial or geographic nature of the pattern to take form, which permits visual inspection and preliminary logical associations and inferences to be made. Through description, the decision maker can effectively "read" the environmental situation surrounding a given problem and assess its status relative to the goals and objects that motivate the decision process.

The prescriptive abilities of a GIS make an important contribution during the goal setting, plan generation, evaluation, and selection phases of the decision process. Prescription, as used here, implies an ability contained within the GIS to direct the analytic manipulation of data in a controlled and structured

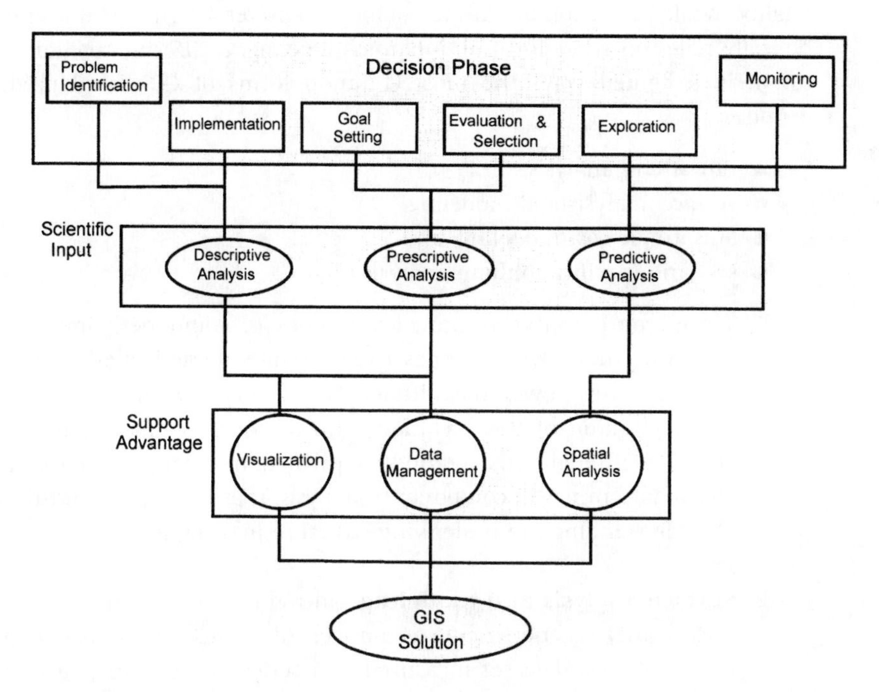

Figure 3.7 GIS Support Functions

way. Through the data manipulation process, guided by human judgment, knowledge, and experience, patterns in data emerge that connect a problem or subproblem to a "realization" that defines some aspect of the problem's solution. The realization produced by this process is an abstraction that:

1. explains a sampling of a possible solution space,
2. illuminates a desired condition, or
3. defines a narrowing down of alternatives to an optimal one, pointing the decision maker in the direction of what is "realized" in the pattern.

Webster (1994) characterizes prescription as an optimization problem that centers around a search for a solution space in a manner that allows alternatives to emerge. In GIS data processing, this search process is typified by Boolean overlay operations and the task of minimizing constraints or maximizing benefits through goal weighing. Application areas commonly employing this technique include suitability mapping, optimal site/characteristics studies, and location/site selection analysis.

Lastly, GIS provides a predictive input to the decision-making process. Although the role of GIS and predictive modeling is unclear, the analytical capabilities of a GIS, when combined with human judgment and reasoning, make certain logical assertions based on the data possible. Examples of this type of prediction include habitat models that relate environmental variables in a GIS in a manner that forms an expression that reflects a prediction of habitat quality. While prediction implies the ability to answer a "what if" question that typically relies on some form of dynamic modeling, a GIS can facilitate predication in a limited way. The most common forms of GIS-based prediction include:

- time-trend analysis,
- surface analysis and modeling,
- anisotropic cost modeling, and
- scenario specification and analysis.

1. **Time-trend analysis**—provides a series of component images, each expressing underlying themes (trends, shifts, periodicities) in the data of successively lower magnitude. An excellent example detailing the GIS application of this technique can be found in Eastman and Fulk (1993). Although this method of prediction is largely an outgrowth of standard principal components analysis, it provides a powerful means of understanding the underlying structure in a sequence of spatial data sets.
2. **Surface analysis and modeling**—introduces the techniques of trend surface analysis and Kriging as a means of predicting the spatial pattern exhibited by a data set measured at a series of discrete (point or site-

specific) locations. Both trend surface analysis and Kriging are referred to as surface estimation algorithms. These methods strive to fit a set of data points to a surface and permit the spatial representation of a continuous variable from a sample. With this estimated surface, the pattern of that variable can be displayed for visual inspection or applied in further analytic operations.

3. **Anisotropic modeling**—explains a general approach to representing phenomena that exhibit differing values when measured along axes that follow contrasting directions. The concept of "cost" is fundamental to this form of surface representation, where cost defines a frictional effect that impedes movement over a surface. The map realization produced by this technique defines a distance or proximity surface that explains the influence of friction in terms of grid-cell values. Because anisotropic friction behaves differently depending upon the direction, this modeling technique can find useful application in such diverse areas as oil spill monitoring, fire modeling, and point source pollution modeling (Eastman, 1995).
4. **Scenario specification and analysis**—identifies an activity undertaken during the early stages of designing a simulation and defines a "picture" of some future state of a system under investigation. This picture of a possible future state can be generated in a limited way with a GIS by systematically altering attribute values, adjusting boundaries, or selectively placing or replacing objects or features on the landscape. By so doing, a simple scenario can be produced and the implications of this imagined future condition can be interpreted and evaluated. For example, a decision maker can evaluate the "what if" ramifications of a site selected for a given economic enterprise, a new highway right-of-way, or a change in zoning or other land use designation. This "soft" technique to modeling, however, depends heavily on expert forecasting and judgment to produce a plausible or credible scenario for a projected trend. Nevertheless, it offers an alternative to the more rigorous mathematical models that cannot be easily integrated into a GIS.

Realizing the GIS Solution

While GIS has many capabilities that are relevant to the needs of the environmental decision maker, its application in formal prescriptive analysis and modeling is perhaps the most important and best developed. Prescriptive modeling, however, can result in a lengthy and complicated procedure before the decision maker arrives at the desired product. Prescriptive analysis is further confounded by the fact that it requires a methodology that blends GIS tools with both procedural and substantive knowledge to design a realistic and accurate model. To the decision maker engaged in GIS-based problem solving, the number of things to control may appear to be overwhelming, resulting in a type of paralysis or anxiety. Therefore, in the beginning stages of prescriptive analysis, the decision maker needs to develop a strategy for effective modeling using the GIS. Either of two basic strategies can be followed: a top-down approach or a bottom-up approach.

1. **The Top-Down Method**—Once the problem has been identified and is clearly understood, the top-down strategy divides the problem into smaller, more workable pieces. Each of the various parts can then be treated in stages that gradually add to the general concept of the problem. For each subproblem a submodel is constructed, and more detail is defined until the final expression can be produced that provides the desired response. To illustrate the top-down method, consider the problem of identifying optimal habitat for the reintroduction of an endangered plant species. The goal is to produce a map detailing locations where reintroduction efforts should be focused. Optimal habitat, however, can be divided into a series of smaller subproblems—for example, microclimatic, topographic, edaphic, disturbance, and ecosystem/vegetative components—each requiring treatment of specific variables. For instance, microclimatic optimality may require consideration of the spatial distribution of temperature, solar radiation, and moisture regimes at critical time periods. These descriptions can then be brought together to produce a single expression of one factor that answers one part of the bigger question. By working on each of the smaller aspects of the problem, analysis is greatly simplified. The top-down approach also facilitates the controlled addition of factors when producing the final expression of habitat optimality.
2. **The Bottom-Up Method**—This is the converse of the top-down methodology. Analysis begins with the smaller components of the problem without concern for the "big picture." Each component is analyzed and assembled based on intuition or theory into a large expression of the problem. While this approach may not be universally applicable as a reliable problem-solving strategy, it can be applied in situations where evidence of the larger problem is unclear, or where there is a lack of theory or understanding relative to the problem. With the bottom-up approach, modeling builds upward toward an answer based on trial-and-error logic. As a demonstration of the bottom-up approach, consider the problem a city may encounter when attempting to zone recently annexed land. The unspecified nature of the zoning problem encourages a degree of experimentation and intuition. Although which land uses to consider and where they should be located may ultimately be a political decision, at the outset of the zoning process economic, environmental, and other factors dominate. For example, there may be an initial

decision to explore opportunities for residential development. An analysis can be formulated to derive an expression of land areas that may be suitable for residential uses. This smaller problem, defined as a residential suitability map, can be reviewed and adjusted as necessary. Other displays of land use opportunity can be produced and examined in a similar fashion. Eventually, by generating a series of opportunity surfaces, a pattern of possible land use emerges. By reviewing, adjusting, and combining these initial displays to reflect other constraining influences, the final description of one possible zoning scheme can be created.

Whether a top-down or bottom-up strategy is selected, the goal is to produce a solution based on the combination of data in the GIS following a logical and testable methodology. Therefore, the data required to provide a solution must be known, its constraints understood, and its role or place in the modeling process specified. As we discussed previously, prescriptive analysis defines the twofold process of (1) sampling a solution space by forming alternatives and (2) narrowing down the set of alternatives to arrive at the most "optimal" solution. Common applications of this modeling problem include site selection, suitability assessment, and constraint mapping.

In general, the decision problem characterizing these applications can be divided into four categories that permit a more formal method of data combination to emerge. Essentially, environmental decisions following a prescriptive design can be explained as one of the following:

1. Single-Criterion–Single-Objective Problem
2. Single-Criterion–Multi-Objective Problem
3. Multi-Criteria–Single-Objective Problem
4. Multi-Criteria–Multi-Objective Problem

A single-criterion–single-objective problem describes the situation where one factor or variable influences the attractiveness of a specific alternative. The objective in this example is a single goal or information product revealing the answer to a question with only one decision point. For example, a land use planner in a rural community may need to know whether septic systems are feasible in a newly annexed section of the planning area. Based on professional judgment, depth to water table may be considered the single most important environmental factor and the one that poses the greatest constraint to effective performance of a septic system. Therefore, a simple decision rule comparing depth to water table against the proposed design of the subdivision can be created to examine the appropriateness of a decision to encourage the use of septic systems in this area.

A problem of this type can be expanded to a single-objective–multi-criteria problem, recognizing that in many situations more than one variable acts as a constraint or limiting factor in a decision. In such a situation, the specific objective defines a more complex functional relationship of the general form

$$O_b = f\left(C_1, C_2, C_3, \ldots, C_n\right). \qquad (3.1)$$

To illustrate this type of relationship, consider the familiar example of developmental suitability assessment. In this example, suitability for a specific type of land use will be a function of several key landscape variables. Through the combination of these variables, an expression will emerge that identifies geographic locations where landscape conditions will support the desired land use. Addressing problems like these frequently involves the development of complicated rating and weighing schemes to insure that the relative importance of each criterion is maintained throughout the analysis.

Often more than one objective is involved in the decision problem and, as is common in environmental decision making, those objectives may be complementary or conflicting. For example, a community's general plan may include goals to preserve open space, reduce vehicular traffic, provide moderate-income housing, and protect endangered habitats. Satisfying several objectives simultaneously has been a long-standing challenge in the decision sciences, and several approaches to problems of this type have been introduced (Carver, 1991; Jankowski, 1995; Voogd, 1983). At issue in multi-objective–multi-criteria decision making is the question of prioritizing objectives under conflict and the well-recognized problems associated with the application of human judgment in ranking, weighing, or rating objectives and criteria (Lein, 1990; Lein, 1993).

In theory, several objectives can be satisfied by a single criterion. For example, a community in central California may have established a series of goals to limit growth, preserve open space, reduce automobile traffic, enhance rural character, and increase recreation access with respect to a specific area of undeveloped land within the city limits. These goals may be achieved by simply generating a distance buffer set to a specific size from an active earthquake fault or other hazard that delineates an area where development should be restricted due to the nature of the risk. Finding parity with respect to a multi-objective–multi-criteria problem is more involved, yet follows the same logical steps: (1) structure a clearly defined problem to form a testable decision rule, (2) evaluate the risks and uncertainty inherent to the decision rule, and (3) apply the rule to assess the results it produces. The principles involved in developing GIS decision rules are discussed in the next section.

Building GIS Decision Rules

A decision rule is nothing more than a procedure to guide the combination of selected criteria from the GIS database and evaluate criteria relative to the disposition of a specific problem or task. A decision rule may, therefore, involve simply setting a threshold value during a search or identifying a specific attribute, or it may describe a more complicated set of operations requiring the comparison of several multi-criteria evaluations (Eastman, Kyem, Toledano, and Jin, 1993). A decision rule generally explains a fixed set of GIS procedures and operations for combining criteria into a derivative product based on the command syntax of the GIS software. In this context, decision rules are the mechanism whereby the data in the GIS database is transformed into meaningful information—something that conveys a specific concept, theme, or new variable to the decision maker.

The process of constructing decision rules rests at the heart of all GIS operations and may be no more complicated than extracting from an existing soils data set housed in the database all the soils in the file that are classified as "sandy loam." The resulting answer, represented as a file, presents that product as a new data set in the GIS for display or further analysis. Conversely, building a decision rule may require developing a sequence of steps and operations to produce a composite index from a set of variables in the GIS database, identify an optimal site, or delineate an area based on a set of evaluative criteria.

From the examples given above, two fundamental types of decision rules applicable to GIS data processing can be identified: (1) those that involve classification and (2) those that require selection (Eastman et al., 1993).

1. **Classification Rules**—employ specific routines in the GIS to group, cluster, or rename objects and attributes into homogeneous (mutually exclusive and exhaustive) categories. A simple example may involve a land use data set where the decision rule requires a separation of land into urban and nonurban categories. Land use attributes in the data set may include categories such as residential, commercial, and industrial, among others. Using a reassignment operation, those classes considered strictly descriptive of urban uses are renamed or assigned into the new urban class. Other categories not in that class are, by default, included in the nonurban category. In other situations, statistical procedures such as cluster analysis or discriminant analysis may be used to group attributes into new class structures.
2. **Selection Rules**—are based on exclusion principles and the application of Boolean operators and operands to organize and control the combination of data attributes in the GIS. In general, selection rules are designed to yield an answer to the generic question "*X* is *a*" where (*X*) defines a spatial location and (*a*) explains a quality or condition of *X* that selects it from all other locations in the data set. An illustrative example might include an agricultural suitability problem where decision makers are interested in finding the best places to encourage cultivation of a particular type of crop. All evaluative criteria in the GIS database are cate-

gorized relative to their suitability for crop type (*a*). Using Boolean operators, data files are then combined to yield a final product that identifies all locations where crop (*a*) can be cultivated.

Boolean Logic

Because it is an electronic device, the computer can recognize only two states: on and off. Therefore, data stored and manipulated by a computer must be represented by a coding system that conforms to these two states. The method developed to facilitate data storage and manipulation employs a binary code that has only two characters: 1 and 0. This binary code also defines a simple two-valued logic system where a quality or condition must conform to one and only one of the two possible states. In GIS data analysis a mechanism is required to guide the logical combination of data given the constraints imposed by this rigid two-value system. This is the primary role played by Boolean logic and the underlying rationale for using Boolean operations for reasoning with spatial data in a GIS (Bonham-Carter, 1995; Burrough, 1986).

Boolean logic and its formal expression as Boolean algebra use the logical operators **AND**, **OR**, **XOR**, and **NOT** to determine whether a given condition is true or false (1 or 0) (Table 3.2). Thus, in the crop suitability example given above, the region of interest explains areas that are suitable for crop (*a*) or not (1 or 0). Deriving this suitability surface requires a series of data operations to produce a series of binary realizations that can be logically combined using the appropriate Boolean operator. To illustrate this procedure, consider the crop suitability problem and assume that suitability is a function of soil type, slope, proximity to market, and existing land use. Each of these layers in the database must be reclassified into a simple binary expression that relates to each vari-

Table 3.2 GIS Logical Operators

OPERATOR	EXPLANATION
EQ	Equal to
NE	Not equal to
GE	Greater than or equal to
LE	Less than or equal to
GT	Greater than
LT	Less than
AND	Selects feature if both evaluating conditions are true
OR	Selects feature only if one condition is true
XOR	Selects feature if one and only one condition is true
NOT	Selects one and only one feature, excluding all others

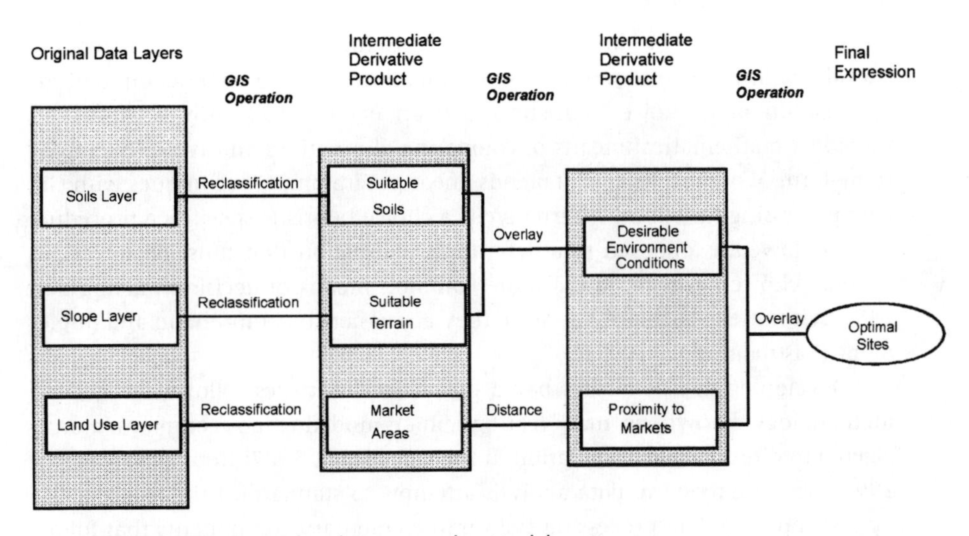

Figure 3.8 Generalized Cartographic Modeling Process

able's suitability to the crop in question. Thus, the soil types in the soil layer must be separated into two groups: those that are suitable for the production of crop (*a*) and those that are not. With the remaining layers represented in a similar binary fashion, logical combinations of these layers can begin. Logical combination employs a Boolean operator to establish a test that evaluates the condition of the data and returns the correct answer. In a raster GIS database, implementing Boolean logic employs the arithmetic operands of addition, subtraction, multiplication, and division to invoke the appropriate data manipulation using the selected layers. In the crop suitability example, the process follows as the combination and simplification of data layers that eventually yields the correct response. This process, illustrated in Figure 3.8, details the use of multiplication to invoke the Boolean **AND** operator and reveals the steps followed in producing the final suitability layer. The steps illustrated in Figure 3.8 also illustrate the general process of cartographic (map) modeling that is central to GIS data analysis and data manipulation (Tomlin, 1990). The appeal of this approach is its simplicity and its close logical association to manual techniques involving the interpretation of transparent map overlays. However, Boolean combinations assume that the criteria involved are of equal importance (Bonham-Carter, 1995).

Refining the Decision Rule

Critical to the task of developing a decision rule is the creation of a method for "deciding." Deciding, in this context, explains the approach that will be followed to reach a solution to the problem. It includes selecting the data that will be used and deciding how the data is to be classified and how the various layers will be combined to produce the derivative product. Two general approaches to "deciding" have been identified: (1) formulation of a choice function, and (2) formulation of a choice heuristic (Eastman et al., 1993). Choice functions provide a mathematical means of comparing alternatives and typically employ some form of optimization that blends linear programming techniques with GIS data processing functions. Alternatively, a choice heuristic specifies a procedure to be followed as opposed to a deterministic function that must be solved. In general, choice heuristics are a more common means of decision making in a GIS environment, primarily because they are easier to comprehend and implement (Eastman et al., 1993).

Developing decision rules based on choice heuristics follows the general methodology known as map (cartographic) modeling and employs a map algebra to direct data manipulation in the GIS (Berry, 1987; Berry, 1991; Berry, 1995). This approach to data analysis attempts to standardize the use of a GIS by decomposing data processing tasks into elementary components that follow a logical, language-driven sequence. This language is a simple modeling language based on natural language concepts that specify analogous data processing routines within the GIS. For example, environmental planners are well acquainted with the idea of representing and manipulating data in the form of maps printed as single-factor transparencies. To provide an operation analogous to the stacking of transparent factor maps, a GIS contains an **OVERLAY** function that provides several options for completing the superimposition of spatial data recorded in digital form. In this example, **OVERLAY** is a processing command and represents an algorithmic equivalent to the manual procedure of stacking one map on top of another. The power of this approach is that the function specified by the language command maintains a logical connection to the manual method. This enables the decision maker to form a clearer picture of the data manipulation process and maintain a link to how the process unfolds and how the stages in an overall methodology are performed. A selection of generic map modeling commands is given in Table 3.3.

Combining map modeling commands into a heuristic demands a sound understanding of the information required and how the data fits into a methodology that will yield the desired end product. For instance, suppose information is needed to depict a region's slope stability. To yield this final product, a decision must first be made as to how slope stability will be expressed on the final derivative map. This initial stage in crafting the decision rule recalls the familiar problem of classification. Although it may seem obvious, it is a critical step in the process since the final product must effectively communicate the theme without overgeneralizing. Likewise, the decision maker does not want to be overwhelmed by the information. Therefore, a classification is sought that effectively communicates, yet still enables a meaningful pattern to emerge from the data.

Once a suitable method of classification has been selected, the factors (vari-

Table 3.3 Sample Map Modeling Commands

TERM	EXPLANATION
AREA	Measures areas associated with data
CLUSTER	Performs cluster analysis on data set
COVER	Imposes layer on top of another
DISTANCE	Measures Euclidean distance over layer
EULER	Determines shape or form of features
EXTRACT	Extracts values from one overlay to another
FILTER	Performs data smoothing
GROUP	Determines contiguous groups in data set
HISTOGRAM	Computes frequency histogram for layer
INTERPOL	Interpolates continuous surface from point data
MIN/MAX	Finds extremes in the data
OVERLAY	Performs Boolean combinations between two layers
PATHWAY	Finds shortest path through a network
QUERY	Extracts irregularly shaped window based on target attribute value
RECLASS	Classifies or reclassifies layer into new categories
RESAMPLE	Fits the surface to a grid system
SURFACE	Performs slope and aspect calculations
TREND	Conducts trend surface analysis
VIEW	Performs viewshed (intervisibility) analysis
WINDOW	Extracts a subimage and saves it

ables) that influence slope stability must be identified in the database. Identifying the relevant criteria draws on knowledge of the problem under investigation and typically requires a judgment concerning which factors to include and why. For purposes of illustration, it is assumed that slope stability is most heavily influenced by:

1. percent of slope (S),
2. land cover (L),
3. bedrock geology (G),
4. soil conditions (K), and
5. drainage (D).

After the appropriate criteria have been selected, the next question to consider is how the selected criteria should be combined to form a credible expression of slope stability. Initially, the hypothesis might be offered that slope stability (SS) resolves to a function of the form

$$SS = f(S + L + G + K + D). \tag{3.2}$$

The above relation implies that the pattern of slope stability can be revealed through the linear combination of the factors *S, L, G, K,* and *D*. Of course, this assumes that the selected criteria participate equally in determining stability. In reality, each factor may exert a different influence, which suggests that the method of combination chosen will also involve weighing factors relative to their importance with respect to slope stability.

This consideration raises the question of choice. Specifically, to implement the decision rule, a choice procedure is needed to guide the process, and the decision maker must select either a choice function based on a recognized formula for calculating stability, or a choice heuristic based on expert knowledge and judgment. In this example, the use of a choice heuristic to determine slope stability is demonstrated. At this point, developing the choice heuristic calls for the ranking or rating of each category describing the criteria selected. One type of index weighing employs factor maps in binary form, where each map defines a single weight. Combining the evidence to yield the desired information product involves multiplying each map by its weight and then summing the maps to be combined, normalized by the sum of the weights. The result produces values (V) between 0 and 1 according to the relation

$$V = \sum(W_i * C)/\sum W_i \tag{3.3}$$

where W_i is the weight assigned to the *i*th map and C is the observed value on the map, which will be either 0 or 1.

When the map categories on each input map are given differing scores, the weighted score (S) for a polygon or grid cell is determined by

$$S = \sum S_i W_i / \sum W_i. \tag{3.4}$$

In this example, Equation 3.3 or 3.4 becomes the choice heuristic and the decision maker's primary task is to assemble the needed data inputs, prepare them for treatment in the choice heuristic, and implement the decision rule that will produce the combination of factors that yields the desired expression of slope stability. Producing this explanation, therefore, relies on the application of the appropriate GIS functions to transform the raw data into the format necessary to execute the selection process and then draw on the GIS to continue the transformation until the desired answer is complete.

Coping with Error and Uncertainty

Rarely, if ever, is the information or data used by a decision maker perfect. Error is an inherent feature of all measured phenomena, and when data is presented in the form of maps and entered into a machine environment, the nature of error and its impact on the correctness of a decision is often masked by the intricacies of the techniques and technologies involved. Thus, there is a tendency to assume that because data resides in digital form and is housed within an expensive array of computer equipment it must be correct. Recent research on the topic of error and GIS reminds us that this assumption is far from valid (Walsh et al., 1987; Dunn, Harrison, and White, 1990; Lodwick et al., 1990).

Early in Chapter 2 error and uncertainty were discussed with respect to the decision-making process. Here, that discussion is placed in a geographic context and linked directly to the application of GIS as a decision tool with a focus on the data and information contained in the system. The common sources of error in a GIS have been noted by Burrough (1986) and examined in detail by Thapa and Bossler (1992). Three general categories of error can be defined:

1. obvious sources of error,
2. error due to natural variation, and
3. machine error.

Obvious sources of error (the first category) are frequently overlooked when designing the initial GIS database. Issues such as the age of the data (maps), aerial coverage, map scale, density of observations, thematic relevance and content, format, accessibility, and cost fall within this category. To a large degree, these factors combined to influence the "appropriateness" of data to serve as input to both the GIS database and the decision-making process. "Old" data may not contain certain critical features or reflect the most up-to-date description of key attributes. Coverage and scale influence the extent to which data is available for the area and at what level of detail. Thematic content and observational density help define the quality of data and how it relates to the problem under investigation. The final considerations of format, accessibility, and cost determine whether the data exists, is usable, and can be acquired.

The second category in Burrough's typology describes error that results from natural variation in the original measurements and speaks to the stochasticity inherent to data collection. This category considers a more data-specific definition of error and data quality, which directs attention to the usability of data and its reliability in an analysis. The principal sources of error identified under this category include factors such as the positional accuracy of data, the accuracy of its thematic content, and data variability introduced by measurement error, method of data collection, or natural (spatial) variability (randomness).

The final category in this three-part classification explains error that develops as a function of machine processing. These unseen errors originate in the software and hardware systems that comprise the GIS and can include numerical errors or inconsistencies that result from rounding, truncation, and limitations imposed on complex mathematical operations; errors associated with complicated "geographical" data processing operations; errors that can be attributed to data classification, generalization, and interpolation methods; and errors generated by software or hardware defects.

In general, inherent error and operational error impart the greatest influence on the accuracy of products generated by a GIS, and their impact can be introduced in a systematic or random manner (Walsh et al., 1987). Inherent error is directly associated with the source documents selected for inclusion in the GIS database and encompasses all aspects of categories 1 and 2 discussed above. Operational error is produced during the process of capturing spatial data and rendering it in the form of a digital file, and through the data manipulation functions of the GIS. In either instance, error can be systematic in that it is consistent and predictable to some degree, and random to the extent that it cannot be aptly characterized by a probability distribution and corrected. Thus, while systematic error can be compensated for, random error introduces a level of uncertainty that might be quantifiable but never completely eliminated. In a GIS setting it is random error that poses the most critical difficulties for effective environmental decision making.

If the assumption is made that data residing in the GIS has been encoded following the strictest of standards, every layer (map) in the database will contain inherent error that can be attributed to the projection used on the source map and how the source map was drawn and symbolized (Vitek et al., 1984). In addition, operational error can be added during data entry, manipulation, selection, and combination within the GIS. Consequently, the decision maker needs information regarding spatial database accuracy and an expression of data confidence that can temper its application in problem solving. One means of acquiring confidence in the spatial data component of the GIS is through a formal and rigorous system of error assessment (Dunn, Harrison, and White, 1990; Finn, 1993).

Standard error assessment techniques are typically conducted by comparing attribute values in the database with those from known ground locations. This process of finding the "ground truth" is comparatively straightforward and can provide the decision maker with an expression that effectively communicates the thematic accuracy of the digital files, although not necessarily the precision of their recorded geographic location. Recently, use of GPS (global positioning systems) has been shown to provide a usable method for determining positional accuracy.

Error assessment begins by constructing a spatial sample of locations to anchor the file (map) to the ground comparison process. The sampling strategy selected may follow either a random, a systematic, or a stratified random design. A random sampling strategy requires a set of (X, Y) locations drawn from a representative probability distribution, and although the random sampling technique is generally unbiased, there is no guiding logic to influence the spatial arrangement of the sample points. Consequently, samples may be placed in a manner that produces poor geographic coverage of the region in question and may not correspond well with the spatial characteristics of the data. Selection of a systematic sampling design requires the establishment of a regular grid of points over the site or its representation as a map or aerial photograph to permit pair-wise comparison. While a systematic design insures complete geographic coverage, bias can be introduced through the design and spacing of the sampling grid. To avoid the potential problems associated with either strategy, stratified random sampling techniques are more commonly employed in error assessment.

A stratified random sample is acquired by dividing the area under consideration into a regular grid of cells with each cell representing a fixed sample location. The exact position of each point within the cells is determined by selecting a coordinate location at random and using that value to select a sample site. The resulting sample maintains good geographic coverage, and because actual points within cells are taken at random, the pattern is relatively unbiased.

Once a sampling strategy has been selected, the next major issue in assessing database error is determination of the size (number of locations) of the sample required to approach statistical significance. Here, finding the error in the data set depends upon the degree of error it contains and the level of confidence and precision the decision maker desires. Since the actual error level of the data set is unknown, an estimate of error is needed. The general formula (shown below) for estimating sample size can be employed to arrive at an initial sample size (n).

$$n = \left(z^2\delta^2\right)/\varepsilon^2 \tag{3.5}$$

where

n = sample size
ε = confidence interval
z = standard score of a given interval
δ = standard error of the estimate.

In situations where qualitative (categorical) data are involved, sample size can be determined by using the formula presented by Eastman et al. (1993):

$$n = z^2pq/\varepsilon^2 \tag{3.6}$$

where p equals the desired proportion correct and q represents the value $(1 - p)$.

After derivation of sample size, the next phase in error assessment is sampling the "real world" and the GIS data set according to the selected sampling strategy. Here, conducting the sample may involve physically visiting sites in the field and noting their positional and attributional values at each location in the sample, or interpreting these values from source data that is assumed to be correct (that is, rectified aerial photographs with corresponding geodetic control). Once this information has been collected, the accuracy of the quantitative measurements can be determined by calculating the *RMS* error according to the formula

$$RMS = \sum \left(X_i - t\right)^2 / n \tag{3.7}$$

where t equals the ground value and x its corresponding value in the data file.

When considering the accuracy of categorical data such as soil type, land use, or vegetation class, the method of contingency table analysis is employed. A contingency table is simply a 2 × 2 matrix designed to compare the known values of the sample with the values recorded in the data set. Tallies are made on a point-sample basis, comparing actual values with those recorded in the file. Tabulations are made of the number of cases (observations) that fall into each combination. Then, when completed, error rates, accuracy rates, and confidence levels can be calculated. Detailed examples discussing the design of contingency tables for thematic map accuracy assessment can be found in Jensen (1995) and Campbell (1987).

Understanding the accuracy that may be attributed to the data feeding into the decision-making process is obvious in its importance; however, it assumes new meaning when data of varying levels of accuracy is transformed and manipulated by the GIS. Spatial data is combined, reclassified, and otherwise modified as part of a problem-solving methodology, so the error inherent to that data is also subject to transformation. This accumulation and modification of the error component of spatial data, when processed via the analytic toolbox of the GIS, have been termed the error propagation effect, which has become an active area of research in GIS data analysis (Heuvelink and Burrough, 1989; Lanter and Veregin, 1992; Heuvelink and Burrough, 1993).

In very general terms, error propagation may be defined as the forward change in error introduced through the combination of spatial data sets as determined by a specific cartographic modeling strategy. More specifically, error propagation recognizes that errors in a spatial data set do not disappear when subjected to analytic manipulation. Rather, through manipulation, error can often be increased and thus degrade the quality of a derived information product critical to the needs of a decision maker (Newcomer and Szajgin, 1984). To demonstrate the importance of error propagation, consider the simplified example of a soils map with 78 percent accuracy and a land cover map of 88 percent accuracy. Assuming that error is multiplicative, the combination of these two layers results in a surface showing land use and soil type with 68 percent accuracy. Although the actual mechanics of determining accuracy in the final derivative product are more complicated, this example illustrates the concept of error propagation and underscores the presence of operational and inherent error in a GIS database. To the decision maker this illustration suggests that while GIS can be an essential information technology, the validity and reliability of decisions made using a GIS can be challenged due to the nature of error.

Although research on the error propagation effect is ongoing, several important indices have been introduced to measure the spatial and nonspatial dimensions of error, and a selection of models has been proposed to assist in explaining the manner by which GIS manipulations generate and modify error. Some of these indices and models are examined below. For a more comprehensive treatment of the subject, the reader is referred to Goodchild and Gopal (1989), Veregin (1989), Veregin (1994), and Veregin and Lanter (1995).

Predicting the level of error expected as a consequence of GIS manipulation can be accomplished either through the use of an error propagation function or by applying a Monte Carlo simulation technique. An error propagation function is simply a formula for calculating error for a given GIS function with a known error index. For arithmetic operations involving the addition or subtraction overlay of two raster GIS files with error expressed in terms of Root Mean Square Error (RMS), the RMS of the output file is determined by

$$M = \left(M_x^2 + M_y^2\right). \tag{3.8}$$

If the two files are subjected to overlay operations involving multiplication or division, the RMS error of the output layer is found by

$$M = \left(\left(M_x^2 * Y_2\right) + \left(M_y^2 * X^2\right)\right) \tag{3.9}$$

where X and Y identify values in corresponding cells of the input layers. This implies that error would need to be computed separately for each cell in the raster layer.

Boolean operations, such as the logical **AND** and the logical **OR**, propagate error to the output layer also. For Boolean (logical) operations, error is typically expressed as the proportion of cells assumed to be in error (PCE) in the category subject to overlay. Since Boolean operations require two files as input, the error propagated to the output layer is a function of two conditions: (1) the error defined in the input layers, and (2) the logical operations. For the logical **AND** operations, error on the output layer can be expressed as

$$PCE_{out} = PCE_x + \left(1 - PCE^x\right) * PCE_y. \tag{3.10}$$

The error propagated by the logical **OR** operation is explained by

$$PCE_{out} = PCE_x * PCE_y. \tag{3.11}$$

In certain cases an appropriate error propagation function may not exist or its solution may be too complex to be cost-effective. In these situations a simpler, more general approach using Monte Carlo simulation has been suggested (Openshaw et al., 1991; Fisher, 1991). The Monte Carlo technique employs variates designed to share the same characteristics of error as those describing the spatial data set in question. With these variates, the propagation of error can be simulated following a procedure outlined by Openshaw (1989):

1. Decide what level and type of error characterize each data set as input to the GIS.
2. Replace the observed data by a set of random variables drawn from an appropriate probability distribution designed to represent the uncertainty reflected in the data inputs.
3. Apply a sequence of GIS operations to the data generated in Step 2.
4. Save the results and repeat Steps 2 through 4 *T* times.
5. Compute summary statistics.

An excellent demonstration of this technique is provided by Eastman et al. (1993). Essentially, this artificial error can be added to the original layer by creating a spatial data set with specific error characteristics. The result produces a perturbation that reflects the values cells in the layer could reasonably explain under conditions of error. Employing this artificial (error) layer in a GIS operation will produce a possible outcome that enables the decision maker to gauge the effect of error-induced variability on the analysis or decision.

Although error and uncertainty have always been features of spatial information when represented on maps, and tend to be "facts of life" in all information systems, they ought not be taken for granted (Openshaw, 1989). Similarly, it is important to realize that not all of the uncertainty embedded in the GIS can be attributed to error or the quality of the spatial data comprising the database. A unique source of uncertainty in a GIS originates in the language concepts a decision maker applies in a machine environment—concepts that the machine cannot understand or process. In this context, uncertainty is a function of the imprecision and inexactness of language and reintroduces the concepts of fuzziness and fuzzy logic described in Chapter 2.

Fuzziness and Fuzzy GIS

Earlier, a fuzzy set was defined as a categorization of data or experience that does not have precisely defined boundaries to discriminate between objects that fall within the class it explains. Common language concepts such as tall, short, warm, and near were used to illustrate these fuzzy sets and further demonstrate the ambiguity inherent to language. While humans can generally reason with fuzzy information, when a decision maker whose language and thinking is rich with fuzzy concepts, fuzzy predicates, and fuzzy hedges enters the Boolean world of the GIS, a level of richness and degree of information is often lost or compromised. This loss of information, or the inability to articulate fuzzy concepts in the two-valued logic system of the machine, produces uncertainty.

To illustrate this point, consider the decision maker who requires a map detailing areas of moderately sloping terrain. The problem facing the decision maker is to define the concept "moderately sloping" and express a class of measured slope ranges that conform to that definition. While a given decision maker may decide that slope ranges between 12 and 15 percent are moderately sloping, others involved in the decision may disagree. If a reasonable range of values can be agreed upon, a membership function can be derived to define the class of slopes that are moderately sloping, where all values of slope can be expressed relative to their degree of participation in the concept. By using this definition of slope, values that completely fit the definition (1.00) can be delineated, while those that might be important but are not defined perfectly by the concept are neither lost nor omitted from consideration.

This approach to reasoning contrasts markedly with the traditional "hard" computing common to the present generation of GIS. Currently, the prime considerations in GIS analysis are precision, certainty, and rigor. A GIS based on the model of "soft" computing, however, recognizes that precision and certainty carry a hidden cost: the loss of information. Therefore, whenever possible, computation, reasoning, and decision making should exploit the tolerance for imprecision and uncertainty (Zadeh, 1994).

Adopting a "soft" computing approach to GIS is still a comparatively new idea, but significant progress has been made, particularly with respect to the concept of fuzziness and the techniques of fuzzy reasoning (Robinson, 1988; Wang, Hall, and Subaryono, 1990; Kolias and Voliotis, 1991; Sui, 1992; Banai, 1993; Altman, 1994). The incorporation of fuzzy set theory in GIS-guided decision making introduced two problems to GIS data analysis that are not easily resolved. The first is the issue of representation, recognizing that the database must be able to hold information about a feature whose location or extent is not known precisely (Altman, 1994). The second issue concerns the application of fuzzy concepts in data analysis and manipulation, realizing that the toolbox of the GIS must be enhanced to facilitate the use of imprecise terms and inexact concepts and measurements.

Methods of Representation

Two general approaches for representing fuzzy information in a GIS have been devised. One method, introduced by Wang, Hall, and Subaryono (1990), employs a fuzzy relational data model to store membership functions derived from items contained in the database. According to this schema, each row in the relation consists of an n-tuple plus a membership grade. Data manipulation in this fuzzy model associates a membership grade to a tuple, relying on five fuzzy relational operators to determine membership grades for the tuples generated by common Boolean operators. Though complicated, this process has an advantage in that the fuzzy relational model can be implemented using conventional relational database management systems.

The second method proposed for representing fuzziness in a GIS applies the concept of a fuzzy region to characterize the graded distribution of geographic phenomena. This approach represents space as a two-dimensional matrix. In this matrix the rows and columns define specific geographic coordinates and the cell elements hold the membership value for that location relative to the fuzzy concept being evaluated. The resulting file describes a fuzzy surface that can be modeled in the GIS. Examples of this representational scheme can be found in Sui (1992) and Altman (1994).

Methods of Analysis

The manipulation of fuzzy sets in a GIS has been approached in several ways. For example, Kollias and Voliotis (1991) describe a fuzzy relational soils information system that contains a query language that has been extended to include fuzzy queries. With this system, it becomes possible to create a decision rule with built-in operators tailored to fuzzy data. In a similar fashion, Altman (1994) describes the application of fuzzy operators in raster-based systems. As with traditional Boolean sets, a full complement of fuzzy operators exists for manipulating fuzzy sets that can be imported to the GIS (Table 3.4). The fuzzy operators most critical to GIS analysis are the fuzzy intersection, which relates to the Boolean **AND**; the fuzzy union, which approximates the Boolean **OR**; and the fuzzy complement, which is analogous to the Boolean **NOT** operator.

The method of analysis using these operators proposed by Altman involves evaluating a clause to determine the strength of the relationship among its arguments. This multistep procedure begins by creating a prototype that serves to define the truth states of the predicate for different values of the domain. Next, the intersection of the appropriate metric relationship and the prototype is computed. The membership of the fuzzy subset produced via intersection is projected onto a single value by using the fuzzy maximum operator to "harden" the fuzzy set to reveal and display the answer.

Evaluating GIS Performance

The rapid increase in the adoption of GIS technology by government and private industry suggests that GIS is viewed as a valuable asset—one that can help significantly in decision-making operations. Yet, despite claims made regarding the potential of GIS technology, there is still relatively little information on the impact GIS applications are having on the organizations in which they have been installed (Campbell, 1994; Budic and Giodschalk, 1994). Improved deci-

Table 3.4 Common Fuzzy Operators

Fuzzy complement:	$\mu A(x) = 1 - \mu A(x)$
Fuzzy union:	$\mu A \cup B(x) = \max[\mu A(x), \mu B(x)]$
Fuzzy intersection:	$\mu A \cap B(x) = \min[\mu A(x), \mu B(x)]$
Algebraic product:	$\mu A(x)\mu B(x) = [\mu A(x) * \mu B(x)]$
Algebraic sum:	$\mu A(x) + \mu B(x) = [\mu A(x) + \mu B(x) - \mu A(x) * \mu B(x)]$

sion making and more effective administration are clearly benefits, yet there is a need to better determine the extent to which these benefits are realized in practice and to provide decision makers with a mechanism to systematically evaluate GIS usability.

A comprehensive case study reported by Campbell (1994) provides some useful insight into the question of realizing the benefits of GIS. Although limited to the experiences of local planning authorities in Great Britain, the results of Campbell's study can be applied elsewhere. With respect to issues surrounding GIS utilization research, findings indicated that:

- the majority of GIS applications under development aim to assist operations activities,
- the key application for most systems has been the creation of automated mapping facilities,
- there was comparatively little use of complex spatial analysis functions; rather, GIS was used primarily for query and display purposes,
- developing working applications took considerable lead time due to system customization, data capture, and the organizational bottlenecks encountered when introducing a GIS, and
- GIS implementation typically requires a sustained push, spanning several budgetary cycles.

Based on these findings, it appears that, at present, GIS is underutilized. It also seems apparent that the successful application of GIS in decision making is as much a social and political process as it is a technical one. Therefore, improving GIS utilization is a priority. According to Campbell (1994), GIS utilization may be increased by encouraging:

1. simple applications that produce information that is fundamental to the decision maker,
2. an awareness of the limitations of the organization and its resources,
3. user-directed implementation strategies,
4. stability with regard to the general organizational setting and personnel, and
5. a capacity to adapt to change.

The issue of GIS usability has been addressed by Davies and Medyckyj-Scott (1994). One hundred and fifty nine respondents to a 90-item questionnaire indicated that they felt positively about GIS, and 51 percent of the respondents used their systems daily. According to this sample, consistency was a major factor in defining a system's usability. The survey results suggest that among current systems the GIS considered nearest to the ideal in terms of usability was a noncustomized raster-based GIS operating on a personal computer (PC). In fact, raster-based products were rated as considerably easier to use, more learnable, and better suited to the tasks of the decision makers than their vector counterparts (Davis and Medyckyj-Scott, 1994). In addition to being simple to use, raster-based systems were more likely to be used for complex analysis and modeling tasks. Users polled in the survey indicated that vector systems were used primarily for mapping and inventory storage applications. In general, evidence of underuse, coupled with low usability ratings, implies that usability is typically sacrificed for added functionality. Only 33 percent of the respondents claimed to have performed complex data analyses. Thus, it would appear, based on this sample, that the majority of GIS functionality goes unused and that systems, at present, are used mainly for basic information processing rather than complex decision making.

Moving to the Next Level

Although GIS has demonstrated potential for being useful in environmental decision making, it still has some limitations that account for its underutilization. As with any new technology, there is a "settling in" period during which targeted users must learn to use it, understand it, and, in the case of GIS, think with it. But bearing this in mind, several unresolved issues involving GIS become apparent. First, the present generation of GISs have a limited capability for accommodating decision problems that involve complex modeling or simulations. Secondly, GIS are essentially "dumb." Like any other conventional computer program, GIS software requires guidance from its users. Users must possess knowledge of both the GIS and the specific problem areas to which it is being applied. Such knowledge can be highly variable; some individuals know a great deal about the GIS and how it works but very little about the specific problem area, while others have considerable expertise in a given subject area but only a very basic understanding of the GIS. Thus, there may be a fundamental need to make geographic information systems "smarter." Finally, the very nature of GIS data processing may introduce a level of task redundancy that frustrates its users. Any operation that must be repeated numerous times quickly becomes tiresome. Although it may be possible to customize the GIS to reduce task redundancy, other approaches to spatial data handling and manipulation are possible and should be considered. The following chapters explore opportunities to extend the utility of GIS and move beyond the spatial database it describes.

Environmental Decision Support

Decision support is an information technology that recognizes the capabilities of a computer to provide essential support functions to the decision maker. These support functions include access to data and models in order to recognize, study, and formulate a problem, and provision of specialized analytic routines to assist in evaluating alternatives and testing the appropriateness of a given decision before committing to a specific choice or action. Understanding this technology, therefore, does much to enhance the utility of database systems such as GIS and offers the potential to link data with models to expand the range of problems that can be examined using information technology. This chapter examines the concept of decision support, beginning with an overview of the motivating force behind the decision support approach—prediction. The discussion continues with a review exploring the role of modeling and simulation in decision making. The chapter concludes with a discussion connecting the concepts of prediction, system analysis, and modeling to the design of decision support systems and their application to environmental problem solving.

The Science of What If

Change is an intriguing subject. Recognizing change, managing it, and predicting its form and consequence is of fundamental importance in all the sciences, but takes on special meaning when attention is directed to the environment. As human societies anticipate a more intensive use of Earth's surface to accommodate future population growth and continued economic expansion, the uncertainties surrounding the processes that propel environmental change and their potential impact demand detailed analysis. The concept of change, as it pertains to the environment, has generated widespread interest among environmental scientists and decision makers (Bennett and Estall, 1991; Cocklin et al., 1992; Roberts, 1994), renewing attempts to better understand the process of environmental change in general and human-induced (directed) environmental change in particular (Robinson, 1991; Gallopin, 1991; Clark, 1989). There has been a continual effort to provide more comprehensive and powerful methods of analysis and prediction of change (Farmer and Rycroft, 1991; Jakeman, Beck, and McAleer, 1993; Jeffers, 1982). It is here that the science of what if begins.

Human-directed environmental change is ultimately the product of a human decision-making process; thus, the dynamics of change are intertwined with a complex web of natural and human-centered forces and influences. From one perspective, the study of change is like viewing the past, uncovering and documenting the progression of a phenomenon. However, when focus shifts to the question of decision making, the outcome of a choice takes us well beyond the comfort of the present and asks the decision maker to confront the uncertainty enveloping the future.

Right now there is an unprecedented sense of urgency and widespread concern about the future status of the biosphere and the recognized capacity of human populations to engender change. As society's capacity to change environmental relationships outpaces our sensitivity to those changes, any approach that can explore the potential outcome of an environmental decision has tremendous societal importance, particularly if it can:

- raise questions concerning (certain) development strategies and objectives,
- provide essential insight regarding the relationship between human actions and environmental degradation, and
- assist in setting priorities in response to economic and population growth pressures.

Addressing questions about the future, an activity undertaken by environmental decision makers every day, is frustrating, fascinating, and complicated. Prior to a selection being made from a set of alternatives, an attempt is made to project that choice into some future state. Whether this is accomplished in a rigorous, formal manner or in an entirely qualitative, heuristic fashion, "future casting" cannot be separated from the decision-making process. Indeed, without the capacity to predict the consequences of a decision, there will be little opportunity to mitigate adverse effects or adapt to changing conditions wisely and realistically. Prediction, therefore, becomes the principal goal of the science of what if, and relies heavily upon the development, testing, and application of models and the representation of "real-world" complexity into simpler structures and conceptualizations.

In this context, representation and abstraction serve as the cornerstones of prediction and provide the intellectual foundation for the study of change. Here, representation explains the process whereby features that characterize the problem as it is perceived in the "real world" are translated into a new expression—one that renders the problem easier to comprehend. For example, the problem of habitat encroachment can be understood by identifying key controlling variables that act on or direct the process. Placing these variables into a simple functional form or relationship allows a rudimentary understanding of how this process unfolds. Abstraction formalizes this representation and describes the general task of transferring an "expression of complexity" into a design that permits cause-and-effect relationships to emerge.

Systems Thinking

Presently, the language of systems analysis and the logic of General Systems Theory are the principal intellectual tools guiding the representation and abstraction of complexity (van Gigch, 1991). Indeed, the primary goal of General Systems Theory is to define models or principles that can be applied to any generalized arrangement irrespective of subject area. Therefore, the underlying concepts of systems theory are as applicable to economic and political processes as they are to geologic and environmental processes. Adopting a system's approach to problem analysis, whether directed toward an environmental or a cultural process, focuses on the system as a whole. Such an approach centers analysis on total system performance, even when a change in only one or a select number of elements is contemplated. This is necessary at times because some properties of a system can only be treated from this "holistic" perspective.

The lure of systems analysis is that it enables the structure and behavior of complex interrelationships to be explored (Bennet and Chorley, 1978). Because the real world is complex, the decision maker reacts by attempting to isolate parts of reality, either in fact or in theory, and then strives to explain how these parts operate under simplified conditions (Chorley and Kennedy, 1971; Wilson, 1981). However, as Chorley and Kennedy state, isolated structures, being subjective and artificial portions of reality, must maintain connection with meaningful sections of the real world. Therefore, when applying system concepts, the jargon of systems-speak and the detailed structures that can be created must still be grounded in method.

A system can be defined in several ways. At one level, a system is simply a part of some potential reality where the decision maker is primarily concerned with space-time effects and causal relationships among its parts (Fishwick, 1995). Another definition explains a system as a set of components, together with specific relationships between components and among their states. Such a definition characterizes a system as a set of interrelated elements that function in a complementary manner and permit the identification of unique process, object, and form. Therefore, a system, whether used to describe a city or a drainage basin, is an entity that is composed of at least two elements and that explains a direct or indirect linkage between those elements.

As an analytical device, a system can be defined at varying levels of resolution and complexity (detail). There are two types of systems: (1) abstract systems, whose elements are concepts and whose components have connecting relationships based on assumption, and (2) concrete systems, in which at least two comprising elements are objects. In most applications, systems are composed of variables that describe the condition of some phenomenon. Such variables, according to the language of systems analysis, are termed "state variables" and represent measurable components of the system whose values can change with time (Huggett, 1993). With respect to the concept of change, the functional relationships that form a system facilitate an understanding of dynamic processes. Change becomes a feature of several key attributes that further refine both the notion of change and the concept of a system. Principal among these are the following:

- **System State**—explains the status of the system at a given moment in time expressed as the set of properties it defines. The system state is defined by the values of its state variables at that point in time.
- **System Environment**—a set of elements and their relevant properties that are not part of the system (external) but can influence its state.
- **System Interaction**—describes how or if a system interacts with its environment. A closed system is completely self-contained and explains no interaction with an element contained by the system. An open system defines some level of interaction with its environment.
- **System Event**—explains a change in one or more structural properties of the system over a specified period of time. Under this concept, systems can be explained as:
    1. *a static system,* where no event occurs and no change is realized,
    2. *a dynamic system,* where an event occurs and changes in state are realized over time, and
    3. *a homeostatic system,* which explains a static system whose elements and environment are dynamic, yet retains its state in a changing environment through internal adjustment mechanisms.

Applying these concepts to represent "real-world" events or processes introduces a four-stage methodology that has been explained in detail by Huggett (1980). These four critical phases in applying systems analysis include:

1. **The Lexical Phase**—which is concerned primarily with identifying the components and controlling the variables that comprise the system, and with defining or describing the components as measurable quantities or states. Identification cannot proceed, however, unless a clear hypothesis

has been defined that can be tested by the system of interest, which then requires establishing the boundaries of the system that separate it from the rest of the environment. Once these requirements have been satisfied, the selection of state variables that define the components of the system can begin.

2. **The Parsing Phase**—which defines the relationships between the components of the system. At this stage, the system linkages are resolved in either a verbal or mathematical manner. The attempt here is to explain the functional relationships, dependencies, or causal structures of the system. By parsing the system's relations, a set of rules are established that govern the behavior of the system, which can be used to explain the direction of flow, the nature of feedback, and other processes that determine the internal workings of the system.
3. **The Modeling Phase**—which involves clarifying the operation of the system and refining the rules that define the relationship between system variables and system states. This phase implies the development of a more formal explanation of the interactions hypothesized between components of the system and their measurement. This phase also demands that the model of the system be operationalized or calibrated by assigning values to variables of the system, including the selection of parameters and constants.
4. **The Analysis Phase**—where the system model is solved to produce testable results. Once the model is activated, it and the results it produces undergo rigorous evaluation. Since all models are inadequate or incorrect to some degree, no model achieves absolute identity with the system it represents. Therefore, analysis of the model produces results given an indication of how well the model fits the system design and how well that system mirrors reality.

As analysis progresses through each phase, the system takes form and the problem motivating the system's approach gains clarity. Process, however, requires more careful definition. Thus, once a system has been identified and its components named, their relation to one another and the underlying process that directs system behavior must be resolved. Generally, system relations and the processes that are defined as flows that connect elements of the system can be described in two contrasting ways:

1. **Continuous versus Discrete Systems**
   The terms continuous and discrete, when applied to a system, refer to the nature or behavior of the system as evidenced by change in the system's state with respect to time. A system whose changes in state occur continuously over time are termed continuous systems, whereas a system whose changes in state occur in finite quanta or jumps are called discrete systems.

2. **Stochastic versus Deterministic Systems**
   A deterministic system is one in which each new state is completely determined by the previous state. The system evolves in a completely fixed and conclusive way from one state to another in response to a given stimulus. Stochastic systems contain an element of randomness, which influences its transition from one state to another. Stochastic systems imply that system behavior may be probabilistic in nature and may be explained by a probability function that can approximate the behavior of the system.

System Design and System Modeling

The ultimate goal of the system design process is the formulation of a model that explains a complex set of relationships in a more tractable form. Through the application of systems concepts and the formulation of a system design, the variety describing a "real-world" situation can be reduced to an ordered, structured set of objects or attributes. To the decision maker, this aspect of systems thinking:

- enables the study and analysis of complex processes,
- accommodates hierarchically structured problems,
- facilitates the study of dynamic relationships, and
- provides a framework for prediction and the analysis of change.

However, realization of a model depends upon the decision maker's capacity for:

1. specifying the variables to include in the system,
2. stating the hypothetical relationships among variables comprising the system,
3. providing a subjective determination of the system's structure, and
4. iteratively testing and refining the system model.

A general schema outlining the method of systems analysis is given in Figure 4.1. The culmination of this methodology provides a design that permits the system model to move to the next level—abstraction.

Abstraction was described previously as the general process of transferring a system design into a "working" model (Fishwick, 1995). Therefore, understanding change in the system involves using this model to explore how the system performs. Modeling, serves as the language for describing the behavior of the system at a given level of abstraction, while the model serves as the metaphor. Thus, to model is to abstract from reality a description of a dynamic system (Fishwick, 1995). In this context, modeling becomes a means of thinking about systems that is similar to the use of systems analysis for thinking about complexity.

Given the complexity of environmental systems, the development and use of models is essential to effective environmental decision making. Because envi-

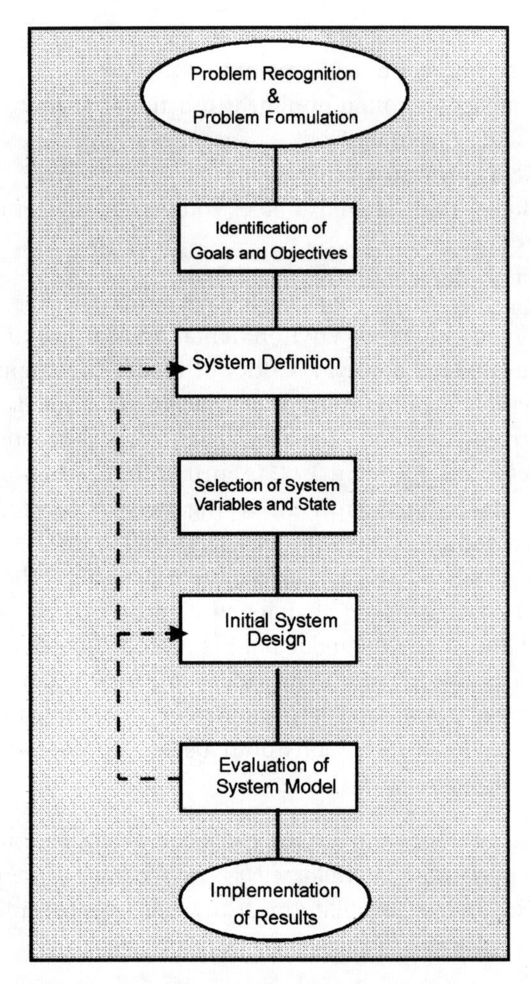

Figure 4.1 General Approach Followed in Applying System Methods

ronmental problems are typically multivariate and dynamic, relying solely on simple calculations and professional judgment is inadequate when evaluating a decision. In any decision problem, the decision maker notices a discrepancy between an existing state and a desired state, has the motivation and potential to reduce this discrepancy, but recognizes that there is more than one possible course of action, any one of which results in an irreversible allocation of resources (Vlek and Wagenaar, 1979). Such decision problems are particularly evident when considering the environmental impact of a decision or the pattern of change a decision may encourage. In both cases, the number of variables involved is likely to be large and the defining system will be too complex to permit intuitive decision making (Gordon, 1985). In these situations, the system involved typifies several characteristics that compel the use of models (Haimes, 1982), including:

1. a large number of decision variables and exogenous and state variables,
2. a large number of components,
3. a complex and often nonlinear functional relationship,
4. risks and uncertainties,
5. a hierarchical structure,
6. multiple, noncommensurable, competing, and often conflicting objectives, and
7. multiple decision makers.

Developing models for environmental decision-making applications is not without frustrations, however. Critics of model-based approaches to environmental decision making argue that models are often inaccurate and overly simplistic representations of reality that distort decisions involving valuable planning resources (Gordon, 1985). Furthermore, when models are employed by decision makers, frequently the wrong model is selected or applied inappropriately. The results obtained through modeling in these cases contribute to significant errors in analysis or in inaccurate conclusions. There are also situations where model use has been abandoned because the decision maker could not understand how to use the model, or because the model did not meet the needs of the problem. Gordon (1985) provides some useful guidance regarding the use of models in environmental decision making. According to Gordon, the following two fundamental conditions must be met to improve the applicability of models in a decision context.

1. A model must simulate the consequences of realistic decisions.
2. A model must be validated as an accurate representation of the real world, and its reliability must be expressed according to defined limits before it is used to support a decision.

Building models to support the decision-making process begins with a problem or question that involves prediction or comparison as part of its answer. While there are no formal rules for developing models, the approach to successful construction of a model proceeds according to the principles of elaboration and enrichment. The model builder begins with a very simple design of the process or problem and attempts to move in an evolutionary fashion toward a more elaborate representation that reflects the complexities of the situation more clearly (Shannon, 1975). Model building, therefore, involves constant interaction and feedback between the real-world system (situation) and the model. Each version of the model is then tested, refined, and validated until a useful approximation is produced. A careful interpretation of the term approximation is essential to understanding the role of models in the science of what if. All models, regardless of their apparent complexity, are approximations containing critical assumptions about the system they represent and the pattern of causality the system exhibits. This simple fact is often lost or forgotten in the rush for results, particularly when models are used by those other than the model's developer(s).

In general, developers of models need to have four interrelated abilities:

- the ability to analyze a problem,
- the ability to abstract from the problem its essential features,
- the ability to select and modify basic assumptions that characterize the system of interest, and
- the ability to enrich and elaborate the model until its results achieve an acceptable level of confidence.

A number of contrasting model types can be described, including (Hugget, 1993):

1. **Hardware Models**—which take the form of scaled, analog representations of the physical entity,
2. **Conceptual Models**—which take the form of charts, pictures, and diagrams depicting system arrangements and flow,
3. **Mathematical Models**—which take the form of numerical expressions that represent critical aspects of process, physical laws, and measured values and relationships, and
4. **Digital Models**—which explain mathematical models translated into a computer language and encoded for machine execution.

While a model can assume different forms, the digital model (the most abstract) enjoys widespread application in environmental analysis and decision making. Taking the form of a computer program, the digital model is a very flexible device in terms of both its design and its ease of use. Digital models also have an advantage in that they provide a platform that: (1) shares the same machine environment as a digital database, (2) can be logically and physically connected to related software systems to form a decision support environment, and (3) facilitates simulation.

Simulation as Information Technology

Simulation explains the process of designing a model of a real system and conducting experiments with this model for the purpose of either: (1) understanding the behavior of the system, or (2) evaluating various strategies for the operation of the system (Shannon, 1975). As a process, simulation describes three tightly coupled activities:

1. **Model Design**—which includes problem formulation, system definition, and model definition and testing,
2. **Model Execution**—which involves calibration, scenario specification and selection, and simulation, and
3. **Execution Analysis**—which describes interpretation of model results and implementation of the results.

As a methodology, simulation offers an experimental environment that enables decision makers to: (1) describe the behavior of systems, (2) construct theories and test hypotheses that account for the observed behavior in a system, and (3) use these theories to predict future system behavior (Shannon, 1975).

Experimentation is the central concept that connects simulation to the larger process of environmental decision making and the goals of effective decision support. While the objective of simulation is to formulate a model that incorporates sufficient reality and detail of a system so that inferences made relative to the model can be extrapolated back to the "real world," placing the model in the context of a decision problem narrows the scope of what a given model may actually achieve. In this sense, a computer simulation experiment is simply a plan or procedure for acquiring a quantity of information. To illustrate this point, consider the urban planner who must understand future population growth in the region and how the possible change in population will impact the demand for land. The experiment that produces this information demands a design that includes developing or selecting the appropriate model and specifying how that model will be applied. With such a plan, meaningful information that satisfies the information needs of the decision maker can be obtained from the experiment. In this context, all models define a level of knowledge: knowledge of process, structure, causation, and pattern. They also reflect our current understanding of the system or the phenomena they represent. Whether it is the dynamics of the atmosphere or the factors that influence the location of a solid waste landfill or industrial facility, models are grounded in theory. Models also draw upon data in the form of variables and parameters, and in their execution, models transform this data into more meaningful information. When all these features of a model are brought together, the simulation model can be regarded as part of the background information on the system of interest, condensing existing knowledge of the structure and the dynamics of process in an organized and precise way (Frenkiel and Goodall, 1978).

The uniqueness of this knowledge and its nature accounts for the use of simulation models as research and educational tools. It also explains why models can play a key role in decision making. Models as information enable decision makers to explore possible future conditions without requiring expert knowledge of the subject or system involved. Furthermore, models educate the decision maker relative to the processes involved, the representation of the system, and the measurement of its components and their dynamics. Deriving meaningful information via a simulation model, however, is not automatic. As with any method of information representation and manipulation, uncertainty and error exist and need to be understood.

In a simulation experiment several important phases of analysis precede the actual implementation of the model as an aid to decision making. These phases concentrate attention on the model itself and its appropriateness first as an abstraction of a "real-world" process and secondly as the correct solution given the nature of the decision problem. Five phases critical to the design of a simulation experiment and the application of a model of information technology can be noted: (1) validation and verification, (2) sensitivity analysis, (3) experimental design, (4) scenario specification, and (5) data preparation (Neelamkavil, 1986; Hannon and Ruth, 1994).

Validation and Verification

Validation is an attempt to express the level of confidence the user of a model can enjoy relative to any inference drawn from the simulation. Because a model is essentially a theory describing the structure and interrelationships of an observable phenomenon, the focus of validation is not on whether the model is a "true" representation of the actual system. Rather, validation sets out to establish the "correctness" of the insights gained from the simulation experiment, and correctness is a matter of degree. Through validation, the operational utility of the model can be determined. This has important implications in the application of the model to decision making since computer (digital) models have an air of credibility that makes them easy to believe. Unfortunately, there is no simple test to establish model validity. Instead, validation consists of a series of tests or trials designed to build confidence in the model. These tests include:

- **Face Validity**—to determine if the results "appear" correct, and
- **Assumption Testing**—to determine if the assumptions can be supported.

Model validation can be divided into a three-stage evaluation process (Fishman and Kivat, 1967) consisting of:

1. **Verification**—insuring that the model behaves as the user intends,
2. **Validation**—testing the agreement between the behavior of the model and the behavior of the real system, and
3. **Problem Analysis**—specifying the correctness of interpretations made with information generated by the model.

Sensitivity Analysis

As its name implies, sensitivity analysis sets out to evaluate how well a model responds to extreme fluctuations in the variables and parameters used to initialize and execute it during a simulation. As a test, sensitivity analysis consists of systematically varying the values of the parameters and/or the input variables over a range of known extreme values. The analyst can then observe the effect of these extreme values on the model's performance and its results. If very little deviation is noted in the model's output, then the model is said to be insensitive to variations in that parameter. The value of such testing is threefold. Initially, sensitivity testing can reveal limitations in the model relative to the values used to parameterize and drive it during a simulation. This can provide important insight into how far-ranging data can be before the model's reliability is severely compromised. Secondly, sensitivity testing can provide an indication of the impact of a decision made under conditions of extreme environmental changes. Lastly, testing a model's sensitivity can offer valuable clues to possible future modification or improvements in its design.

Experimental Design

An experimental design selects a specific approach for gathering the information required to draw valid inferences from the model. Three essential steps in selecting an appropriate experimental design are noted by Shannon (1975):

- determine the experimental design criteria,
- synthesize the experimental model, and
- compare the model to standard experimental designs and select the optimal design.

As in any other design problem, the simulation experiment must be approached systematically, and careful consideration must be given to the criteria chosen to form the final design. Among these criteria are:

- the number of input variables (factors) to be varied,
- the number of values to be used for each factor, and
- the number of measurements of the output variable (response) to be taken.

Scenario Specification

The concept of a scenario has been defined in several different ways (Kopik et al., 1982; Wilson, 1978; Ross, 1989; Heydinger and Zenter, 1983). Perhaps the most general definition of a scenario explains the concept as an exploration of an alternative future. More exacting definitions characterize a scenario as a possible sequence of events constructed for the purpose of focusing attention on key process and critical decision points (Wilson, 1978; Robs, 1989). Under a given set of assumptions, a scenario expresses the conditions and events that constitute the "what if" situation decision makers are interested in evaluating. As an integral feature of simulation modeling, the scenario establishes the background against which predictions are made and explains the details of a given causal path that provides a structured set of outcomes for analysis. For example, in an attempt to model various outcomes of the possible growth-inducing effects related to a proposed new transportation route, scenarios can be constructed to evaluate various design and location questions, origin and destination patterns, levels of spatial interaction, and changes in economic conditions. Wilson (1978) specifies the following distinguishing characteristics of a scenario developed to assist the simulation process.

1. Scenarios are hypothetical. Since the future is unknowable, the best that can be achieved is an alternative possible future state of the system.
2. Scenarios are essentially an outline seeking only to (a) identify the major determinants that might cause the future to evolve in one specific direction rather than another and (b) sketch in the principal consequences of a given chain of events.
3. Scenarios should be multifaceted, holistic, and dynamic in their approach to the future state of the system under investigation. A clearly defined scenario should be capable of representing multivariate, interacting flow processes and combining (when appropriate) the critical events, variables, and effects that explain the decision problem.

Developing a credible scenario to guide the simulation process involves compiling a detailed listing of all phenomena potentially relevant to the problem together with the process and events that influence change in the system. Heydinger and Zenter (1983) offer the following four-stage approach to scenario development that is well suited to the environmental decision-making process.

1. Determine the purpose and time horizon of the scenario.
2. Select the elements to include in the description.
3. Adopt a premise that details the assumption inherent to the scenario.
4. Generate a series of credible scenarios and select a set of the more probable ones for analysis.

The fourth stage of this procedure depends largely on the ability of the decision maker to clearly identify the potentially relevant processes so that those considered incredible, irrelevant, or unlikely can be eliminated. Selecting the set of "credible" scenarios to model typically requires an estimate of likelihood expressed as either a scenario probability or a value derived by means of event tree analysis (Ross, 1989).

Model Preparation

With the scenario(s) specified, the model developed or selected for application can be executed. Executing the model, however, requires preparation of data for input and calibration of the model to insure its proper operation. Data preparation may be comparatively straightforward, requiring variables and parameters that can be taken from easily accessible sources. In other instances, data preparation may involve detailed data-gathering efforts before data files can be assembled and passed to the model. To the decision maker, the critical factors to consider when selecting a model, apart from appropriateness, are data quality, data format, and reliability. However, depending on the model selected for the analysis, issues such as the volume of data required, the size of the data file when assembled, and execution time may influence the feasibility of the simulation exercise. Therefore, while the specifics of a model are central to its applicability as a decision support aid, data requirements and their availability are also important factors to consider when choosing and implementing a computer model (Gordon, 1985).

The range of computer models available for simulation in the areas related to environmental decision making are numerous. When a model is selected, validated, and tested for the application in question, placed into an experimental design, related to a clearly articulated scenario, and merged with the required data, it can fulfill several demands implied by the decision-making process, including:

- the simulation of alternative decision outcomes,
- the evaluation of a set of recommended actions,
- the identification of potential consequences, and
- the testing of different management strategies.

When applied in any one of these capacities, the model and the simulation experiment designed to implement it move the decision-making process squarely into the realm of decision support. In the following sections the concept of decision support and the components of a decision support system (DSS) are explored. To set the stage for this discussion, the range of environmental models that can be placed into a decision support role are reviewed.

Environmental Decision Models

The complexity of environmental systems and the practical difficulties associated with empirical studies of environmental processes have encouraged the widespread development of computer models as an aid to research and education. Nearly every facet of the environment has been examined through the use of models to some degree. Perhaps the most comprehensive summary documenting the range of available environmental models can be found in Jorgensen, Nielsen, and Sorensen (1995).

From an applied perspective, the wealth of computer models that can be called upon to assist various aspects of the decision-making process offers a broad spectrum of applications that can be assembled to form a computer decision support system. The range of models available to the decision maker is beyond the scope of this section to adequately address. However, there are specific classes of models that can be examined, and their relative applicability and decision support potential can be described. Three principal categories of application-focused models can be identified: (1) digital process models, (2) spreadsheet systems, and (3) general-purpose simulation programs. The models that fall into these categories provide the decision maker with the critical capacity to simulate key environmental processes. The differences separating the three classes are generally a matter of procedural implementation, complexity, and relative sophistication.

Digital Process Models

Models that comprise this category describe computer simulation models that have been developed to simulate key environmental processes or to explore critical environmental problems. Digital process models typically describe specialized computer programs designed to function as stand-alone systems that require data and parameters placed into formatted input files and that generate their own unique output. Examples of the range of digital process models available include land surface process models, surface and subsurface hydrologic models, ecosystem models, and contaminant fate and transport models. A limited selection of models arranged by these topical headings is provided in Table 4.1. Correct use of any of these models requires an understanding of the environmental system under investigation together with the specific details that control operation of the model. In addition, a certain level of compatibility is assumed between the structure and spatial resolution of the model and the data

Table 4.1 Selected Environmental Models by Subject Area

AIR QUALITY	
CALINE 3	Line-source dispersion model, used to predict carbon monoxide concentrations for transportation applications
CDM 2	Climatological dispersion model, used to determine long-term pollutant concentrations at any ground-level receptor
INPUFF	Gaussian integrated puff model, used to evaluate accidental release of a substance or to model continuous plume
MESOPUFF	Lagrangian model, used to calculate transport, diffusion, and removal of pollutants from multiple point and area sources
PEM	Urban-scale model, used to predict short-term concentrations and deposition fluxes
WATER QUALITY	
CE-QUAL-W2	2-D longitudinal-vertical hydrodynamic and water quality model for water bodies
SELECT	1-D vertical steady-state model for selective withdrawal from a reservoir
STEADY	1-D longitudinal steady-state stream temperature and DO model
SUBSURFACE HYDROLOGY	
MODFLOW	Modular 3-D finite-difference groundwater flow model
MODPATH	Modular finite-difference contaminant transport model
SUTRA	Variably saturated flow and transport model

required to drive the model. Therefore, selection of the appropriate digital process model as an aid to decision making occurs after the information to be gained from the model (that is, the questions the model will answer) is understood and the data required to drive the model is collected.

Spreadsheet Systems

Electronic spreadsheet systems describing software programs that (1) store data as a two-dimensional table, (2) permit calculations with this data, and (3) instantly display results in a variety of graphic formats have become an alternative approach to data analysis as well as an attractive platform for environmental modeling (Gordon, 1989; Hardisty et al., 1993; Cartwright, 1993; Klosterman et al., 1993). As Klosterman et al. state, modern spreadsheet programs provide an extensive array of built-in functions that greatly enhance information display, computational procedures, and simple database management. In addition, spreadsheet systems provide a range of sophisticated commands that can be used to automate repetitive tasks and develop templates and models that can be employed in a wide range of analytic procedures. This point supports the observation made by Cartwright (1993) that spreadsheet systems make the tools of simulation modeling accessible to decision makers in a much broader context.

Although spreadsheet systems were originally designed for financial analysis, they have evolved into a programming environment that has several important advantages as a modeling tool (Cartwright, 1993). These include:

- relative ease of programming,
- comparative ease of modification,
- a transparent design that makes them understandable to the user,
- a structure that allows construction of models that invite "what if" modeling,
- a functionality that provides power and flexibility, and
- a built-in capability for generating graphics.

Several disadvantages associated with spreadsheet systems can also be noted. Central among these are:

- speed—spreadsheet systems are slower and less elegant when compared to models written in conventional programming languages, and
- iteration—iterative processing is difficult to achieve in a spreadsheet format.

In spite of these limitations, a wide assortment of spreadsheets designed for environmental modeling have been introduced to demonstrate the potential of spreadsheet systems and to provide an easily accessible means of studying environmental process. The types of models developed in spreadsheet style include subject areas ranging from demographic analysis and environmental analysis to economic forecasting and decision making (Klosterman et al., 1993).

General-Purpose Simulators

General-purpose simulators define a family of computer languages designed to support simulation modeling. These languages typically provide a syntax that greatly simplifies the representation of elements comprising a system and a structure that enables characterization of process within the defined system. Functions common to the majority of simulation languages include facilities for:

1. generating random variables,
2. managing simulation time,
3. controlling events,
4. collecting and representing data,
5. summarizing data, and
6. formulating and producing output.

Selecting the appropriate general-purpose simulator is influenced by a number of factors that direct their applicability to decision making. Several of the more important ones have been summarized by Graybeal and Pooch (1980) and include consideration of the decision maker's familiarity with the language, the ease with which the language is learned, the complexity of the model it produces, the availability of support, and the inherent flexibility of the language. These considerations provide a basis for evaluating the relative appropriateness of a given simulation language as a modeling environment. However, when considering using a model developed from a general-purpose simulator in a decision support capacity, the following four evaluation criteria need to be considered: (1) support of basic functions, (2) debugging assistance, (3) ease of use, and (4) flexibility.

Of the languages available to the decision maker, those that model continuous systems are of particular importance, three of which are: (1) block-form languages, (2) expression-based languages, and (3) graphic-based languages. A block-form language is constructed in block diagram form. This model is then implemented through a set of connection statements. The common functions accomplished by particular blocks are specified, and the user is not concerned with the details of their development (Graybeal and Pooch, 1980). With an expression-based language the model is implemented in equation form. Creating a model using an expression-based language is similar to programming in a procedural programming language such as Pascal or Fortran. A graphic-based language replaces the procedural statements of an expression-based language with graphic symbols that represent specific modeling conventions. The graphic symbols can be arranged to compose the desired model and "opened" to permit the initialization of variables and parameters along with other conditions that specify the scenario. Environmental modeling using a graphic-based simulation language is demonstrated in Hannon and Ruth (1994).

Regardless of language style, a continuous-system simulation program consists of three sections: (1) an initialization section that establishes the values of state variables that represent the initial conditions of the problem, specifies the values of constants in the model, and assigns parameters of the model such as the time step and analytic time horizon, (2) the main program, which describes the equations of the model that will be used to generate a solution, and (3) a termination section that controls creation and formatting of the model results. When these components are assembled to form a model, they provide a modular structure that can be used to represent a range of complex, dynamic relationships.

Although environmental process models are created to function in the computer as independent routines, the decision maker, when exploring a given environmental question, may require access to several models during the course of an analysis and wish to share data among or between specific simulations. There may also be situations where the output of one model can serve as input to another. While this type of integrated environmental modeling is not uncommon (Brouwer, 1987), a machine environment that can facilitate access to and manipulation of data and models to support decision makers is still comparatively untested. In the following section, the opportunities for this form of decision support are explored and the strategies for creating decision support systems are reviewed.

Decision Support

Designing automated systems to support decision making is not a new idea. The concepts and examples of decision support have been reviewed and described extensively, although from a business management perspective (Vlek et al., 1993; Konsynski, 1992; Lewandowski, 1991; Davis, 1988; Hopple, 1988; Bonczek et al., 1984). The promise of automated decision support and the successful implementation of systems capable of assisting decision makers in corporate planning has encouraged their development in problem areas more closely aligned with environmental management and the needs of the environmental decision maker (Fields and Kim, 1992; Beroggi and Wallace, 1992; Mulder and Corns, 1995; Thomson, 1996). In all of this literature, the underlying rationale for automated decision support remains the same—to supply timely and accurate information to strengthen and improve the effectiveness of the decision-making process. In order to understand this role, we begin by dissecting the concept of decision support and examining the features of a decision support system (DSS).

The motivating force behind the concept of decision support and the development of an information technology designed to aid in the decision process is essentially the goal of helping people make better decisions (Vlek, et al., 1993).

Helping people make better decisions, however, is not a simple task, and, as noted by Vlek et al., it depends on the answers given to the following fundamental questions:

- What is a decision?
- What are better decisions?
- What makes one decision better than another?
- For whom are decision support methods intended?
- What constitutes help in making decisions?
- Who is providing the help?
- Is help really needed?

The decision-making process was examined in detail in Chapter 2. Here, the process is reduced to three global components:

1. acquiring, retrieving, and selecting relevant information,
2. structuring the decision problem to enhance the visibility of the alternatives and their features, and
3. evaluating alternatives for their relative expected attractiveness.

Assuming that the decision-making process can be reduced to these three principal activities, decision support becomes a methodology that can be developed and applied to each one (Vlek et al., 1993). In this context, the underlying assumption and "test" guiding decision support is that methodologically supported decisions should be "better" than those that are unsupported. For example, decision support directed at information acquisition would imply that expert knowledge from persons or documents is provided in an automated setting so that the decision maker has optimal knowledge of the problem environment in which the decision is to be made. The decision maker, armed with sufficient information, would then draw on decision support to direct the selection and ordering of decision-relevant factors in order to insure that a transparent, logical structure of the problem emerges. Then, when a well-structured problem takes form, decision support would guide the evaluation of choice alternatives as specified by the adopted decision model.

Throughout the three stages listed above, the goal of decision support is to channel and direct the decision maker's cognitive processes in order to reduce the uncertainty about what would be an acceptable or "best" course of action (Vlek et al., 1993). Given the fact that human beings, in general, are limited, selective, sequential information processors capable of making comparatively simple, adaptive decisions, but poor at making complex, strategic decisions, the role of decision support becomes clear. Because the quality of unaided decisions is largely dependent on task complexity, there is a tendency toward simplification as the decision task becomes more difficult. Although simplification can lead to effective decisions, simplification can also contribute to systematic errors and defective choices, and result in suboptimal decisions. The role of decision support in the face of complexity is to help overcome cognitive limitations—limitations that become realized as selective perception, selective information retrieval, incomplete or biased problem definitions, and unreliable or inconsistent evaluation of alternatives. Decision support in this capacity functions to extend the decision maker's informational reasoning and evaluation capabilities such that the decision maker operates as a simultaneous information processor (Vlek et al., 1993). Above all, however, effective decision support should be efficient with respect to the context of the decision problem.

Decision support can benefit the decision maker in a broad spectrum of problems (Davis, 1988). Davis divides these situations into five classes based on their relative degree of difficulty (Table 4.2). Overall, decision support appears to be best applied when the decision problem is complex or ill-defined (semi-structured/unstructured). According to the language of decision support, a problem or situation is considered unstructured (ill-defined) when: (1) objectives are difficult to determine or conflicting, (2) alternative actions that might be taken are difficult to determine, and (3) the effect of an alternative on a given outcome is uncertain (Konsynski et al., 1992). In such cases, decision support is targeted at the "structurable" part of the decision problem. The truly unstructurable aspect of a decision problem is usually left for the decision maker to resolve.

Automated Decision Support

Automated decision support defines a computerized system that utilizes knowledge about a particular application area to help decision makers working in that area solve ill-structured problems (Bonczek et al., 1984). To be effective, such systems must possess at least one of the seven basic decision-making abilities listed in Table 4.3 and demonstrate that ability in some stage of the decision-making process. The typical decision support system exhibits a combination of the following capabilities:

Table 4.2 Classification of Decision Problems Suited to DSS Automation (Davis, 1988)

PROBLEM CLASS	TASK DEFINITION
Type O	mechanistic; non-thought-provoking; compliance with limited alternatives; singular objectives; few complications
Type A	use of computer as visual aid; data sorting and data search; automation of routine computations
Type B	issues related to trade-offs; problems with conflicting objectives
Type C	structural complexity; large in size and scope; difficult to visualize
Type D	complex; dynamic; highly qualitative

Table 4.3 Critical Abilities Required of Decision-Making Systems

1. Ability to collect information
2. Ability to formulate models
3. Ability to govern problems
4. Ability to analyze problems
5. Ability to evaluate problems
6. Ability to recognize problems
7. Ability to implement strategy

1. information collection,
2. problem recognition, and
3. analysis.

Information collection manifests itself in a decision support system as the process of gathering data and information from the user of the DSS and from some repository of data pertinent to the application area, usually a database. Problem recognition explains the refinement of the initial definition of the problem through data visualization and discovery techniques embedded in the decision support system, while analysis involves interfacing collected information with a model for the purpose of prediction or explanation. At this point in the discussion, a distinction between decision support systems and information systems in general is needed since the two technologies can have similar objectives.

According to Bonczek et al. (1984), a decision support system is specifically oriented toward a type of information processing activity that constitutes a decision process. Automated support provides an environment that enables the decision maker to manipulate large volumes of data, perform complicated computations, and investigate a variety of factors that would otherwise go unnoticed (Davis, 1988). The decision support system must, therefore, possess application-specific knowledge (Bonczek et al., 1984) such that the automated system:

- supports rather than replaces decision makers,
- can be applied to semistructured or unstructured situations,
- focuses on effectiveness rather than efficiency,
- supports all phases of the decision process,
- employs relevant data or models, and
- facilitates learning.

Based on these characteristics, a decision support system, with its emphasis on decision making, differs from an information system in three major ways:

1. it is designed to support the solution of ill-structured problems,
2. it furnishes its users with a powerful language for problem solving, and
3. it incorporates models directly into its design.

With these three distinctions understood, the components of an automated decision support system can be explained. In general, a decision support system consists of three tightly integrated elements: (1) a language system, (2) a knowledge system, and (3) a problem-solving system, or what have become more commonly referred to as a dialog generation and management system, a database management system, and a model-based management system, respectively (Bonczek et al., 1984; Davis, 1988; Sage, 1991). The relationship between these elements is illustrated from the DSS user's perspective in Figure 4.2. Together, these pieces complete the functional requirements of a DSS as suggested in Figure 4.3.

The language system of a DSS is the total of all linguistic facilities made available to the user of the system. The language system is designed to satisfy the knowledge representation, control, and interface requirements of the DSS.

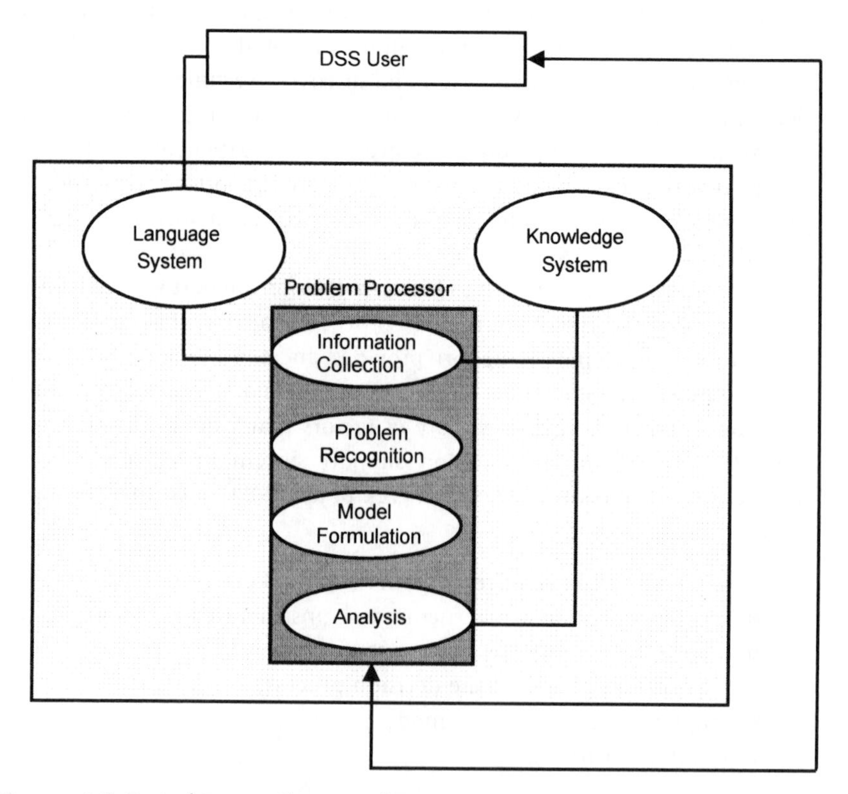

Figure 4.2 Typical Design Elements of a Decision Support System

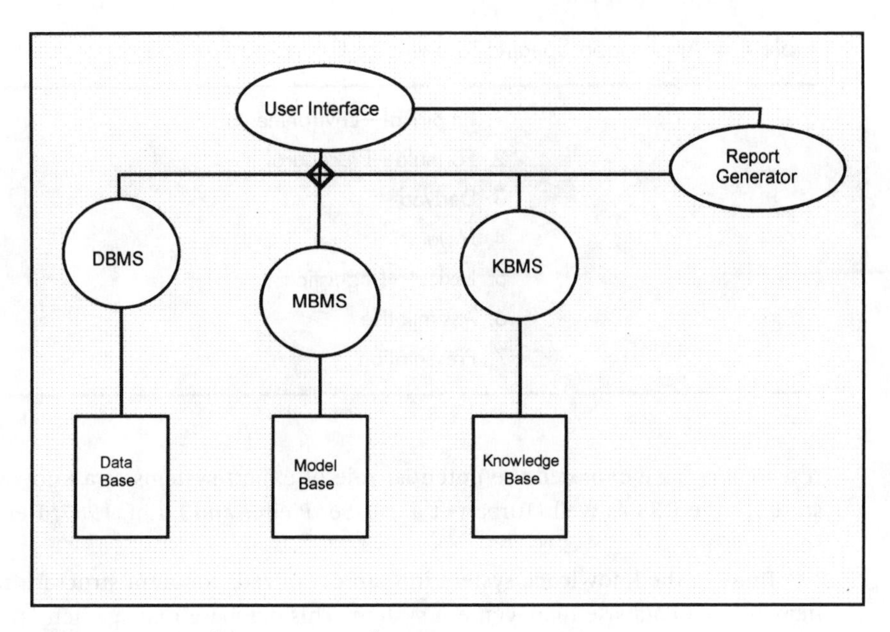

Figure 4.3 Critical Components of a DSS

Typically, language systems direct data retrieval and numerical computation. The language system supports dialog between the system and the user and controls the execution of models housed in the DSS. A range of possible dialog formats can be crafted into the language system, and all are inherently connected to the representational forms employed by the database and model-based management systems (Sage, 1991). In addition to directing data retrieval and model execution, the language system functions to provide a mechanism for maintaining the contents of the DSS knowledge system. Here, the language system operates to assist the review and sensitivity analysis of past judgments and provides partial judgments based on incomplete information (Sage, 1991).

The knowledge system of a DSS represents an organized store of application-specific knowledge about the problem domain in which the DSS has been designed to function. This knowledge can be accessed by the user during consultation with the DSS. Although several distinct types of application knowledge have been identified (Holsapple, 1983; Holsapple and Whinston, 1983), as suggested by Table 4.4, not all are present in the knowledge system of a DSS. Depending on the approach used to construct the DSS, the knowledge system can be represented in the form of a spreadsheet, file, or database. Knowledge stored in the form of a spreadsheet can be used to represent both empirical and formula knowledge. Lexical and other linguistic knowledge can be stored in files and connected using relational data structures. Procedural knowledge can be represented in the form of program command files or as a spreadsheet. Several

Table 4.4 Application-Specific Knowledge Critical to the DSS

1. Empirical—Environmental
2. Formula—Procedural
3. Derived
4. Meta
5. Lexical—Linguistic
6. Assimilative
7. Presentation

researchers have explored the potential role of expert systems as a knowledge store for the DSS as well (Turban et al., 1986; Pfeifer and Luthi, 1987; Henderson, 1987).

Binding the knowledge system into an organized, coherent structure relies heavily on a database management system. This database management system must be capable of working with data that are internal to the decision maker as well as coping with a variety of data structures that permit operations on probabilistic, incomplete, and imprecise data (Sage, 1991).

The problem-processing or model-based management system is the heart of the DSS. The single most important characteristic of the model-based management system is that it enables the decision maker to explore the decision problem via the database by a complement of algorithmic procedures and associated model management protocols (Sage, 1991). Through its ability to acquire and access information, the problem-processing system takes the problem, connects with the appropriate domain knowledge, and then, using its application-specific model base, derives a solution that supports the decision maker in some capacity.

The overriding purpose of the model-based management system as it relates to problem processing is to transform data from the database management system into information that is useful in decision making. Therefore, the model-based management system functions to create, store, access, and manipulate models to satisfy five interrelated decision-making objectives:

1. to facilitate the efficient creation of new models,
2. to support maintenance of a wide array of models that assist problem formulation, analysis, and interpretation in a problem domain,
3. to provide for rapid model access and integration between the actual model and the database,
4. to centralize model management, and
5. to maintain the integrity, consistency, and security of the models selected for inclusion in the DSS.

In a detailed discussion of model-based management systems, Sage (1991) identifies two types of model-processing efforts critical to this component of the DSS: the model-processing-model-based management system, and the decision-processing-model-based management system. According to Sage, the user of the DSS interacts directly with the decision-processing module, while the model-processing component is concerned primarily with model maintenance activities. The range of models that can be included in the model-based management system to support decision making with a DSS fall within three broad categories (Sage, 1991):

1. **Problem Formulation Models**—assist in the identification of problem elements and characteristics. Problem formulation models can include interaction matrices, trees, and structural modeling methods.
2. **Analytical Models**—describe specific system models designed to address a particular aspect of the decision problem. Analytical models can include gaming models, trend and time series models, dynamic simulation models, statistical models, and mathematical programming models.
3. **Interpretive Models**—assist with the evaluation and comparison of alternatives. These evaluation and choice-making models can be based on decision analysis, multiple-attribute utility theory, social judgment theory, or other methods of alternative elimination.

Developing Decision Support Systems

Various strategies have been presented to guide the physical creation of a decision support system. These approaches can be found in Sage (1991), Sprague and Carlson (1982), and Mitta (1986). In many respects, the design considerations of a DSS are similar to those described in relation to the geographic information system and include factors such as:

1. specification of system requirements,
2. conceptual design,
3. logical design and specification,
4. detailed design and testing,
5. implementation, and
6. evaluation and modification.

Although none of these factors should be ignored or discounted, discussion here is limited to the logical design and architectural specification phases of DSS construction.

The most comprehensive description of DSS design is found in Sprague and Carlson (1982). In this treatment of decision support systems, the authors describe a three-part design focus that introduces two specific views of the issues surrounding DSS construction that are of particular relevance to environmental decision support: (1) the builder's view, which concentrates on the technical capabilities of the system, and (2) the toolsmith's view, which details the science and engineering required to create the information technology to support DSS and the specific architectures developed to combine the basic components of the DSS into a functioning system.

Recalling that a DSS consists of database, model-base, and dialog generation and management software components, the builder's view of the design problem considers how to configure these elements to produce the desired level of decision support. The critical capabilities to consider, from a builder's perspective, are listed for each DSS subsystem in Table 4.5. Each capability listed describes a particular level of functionality in the DSS and instructs the developer of the system in which features are central to the goals of producing a useful DSS.

The toolsmith's view of DSS development concerns the specific software engineering needed to make the system a functioning reality. From this perspective, the key aspects of DSS design involve selecting the appropriate "tools" to include in the system to obtain the desired level of decision support. The paramount concerns at this stage of DSS development include:

- selection of compatible software systems,
- software architectures and software interfacing,
- data structures and database models,
- programming support and selection of a programming language,
- software integration and customization,
- functionality testing, and
- DSS module interfacing.

Table 4.5 DSS Functionality Described by Module

COMPONENT	FUNCTIONAL NEEDS
DSS Dialog Generator	handle range of dialog styles; accommodate input from range of devices; present data in assortment of formats; provide flexible support
DSS Database Component	combine variety of data sources; update files quickly; portray logical data structures; manage diverse data with range of functions
DSS Modeling Component	create new models quickly; access and integrate modeling elements; catalog and maintain model base; link models to data sources; manage models with range of functions

Since a DSS consists of three distinct software modules dedicated to the tasks of dialog, database, and model-based management, mechanisms that provide for their functional integration are needed to insure near seamless operation. Complete integration, however, is often difficult to achieve. Six typical problems related to integration are:

- poor response times,
- inability to execute large modules,
- inability to interface the dialog component with the modeling and database components,
- inability of maintenance programmers to understand the software structure,
- high development, operating, and maintenance costs, and
- inability to connect the DSS database with other internal or external databases.

Sprague and Carlson (1982) maintain that effective integration is ultimately a function of the DSS architecture. Therefore, selecting the appropriate architecture is critical and defines a choice made on the basis of several operational criteria including usability, cost, performance, adaptability, and reliability. These five factors can be used by system developers to compare alternative DSS architectures, and when each criterion is satisfied to some level of acceptability, the components of the DSS can be connected with a measure of confidence.

Four principal DSS architectures have been defined, although other configurations are possible (Sprague and Carlson, 1982). These four system architectures—the DSS-Network, the DSS-Bridge, the DSS-Sandwich, and the DSS-Tower—are functional configurations suggested exclusively by their software structure. As such, these designs also represent fairly generic configurations for which operational examples exist. The basic features of these system architectures are reviewed below. For a more comprehensive treatment of these designs, the reader is referred to Sprague and Carlson (1982), and Sage (1991).

The DSS-Network

This configuration takes an adaptive approach to component integration and has been designed primarily to permit different modeling and dialog components to share data and simplify the addition of new, nonhomogeneous components. An important feature of the DSS-Network is that its design facilitates the intermixing of components developed by different groups, at different times, with different programming languages, for different operating environments. The DSS-Network achieves integration through the creation of specific component interfaces. Thus, for each dialog or modeling component contained in the

DSS, there is an interface that takes the form of a communication portal between one component and another. According to this strategy, component interfaces do not communicate directly with one another. Therefore, if one component must be shared by another, the shared component provides for the scheduling and isolation of communication from the component interface. Interfacing in this manner enhances flexibility by combining elements, and enables the addition of independently developed components by virtue of the interface that must be created for each dialog of each modeling element introduced into the DSS. A simple schematic of the DSS-Network is given in Figure 4.4.

The DSS-Bridge

This design creates a unified interface component between (1) the dialog, local modeling, and database components of the DSS, and (2) the shared modeling and database components, if any. This unified interface reduces the number of component interfaces in the DSS (such as those created using the DSS-Network design), but retains the ability to integrate new components into the system. Because the bridge is a centralized and standardized design, all local components and all shared components execute in the same environment. This design feature allows the bridge to perform format conversions and synchronization functions with fewer interfaces. The basic design of the DSS-Bridge is shown in Figure 4.5.

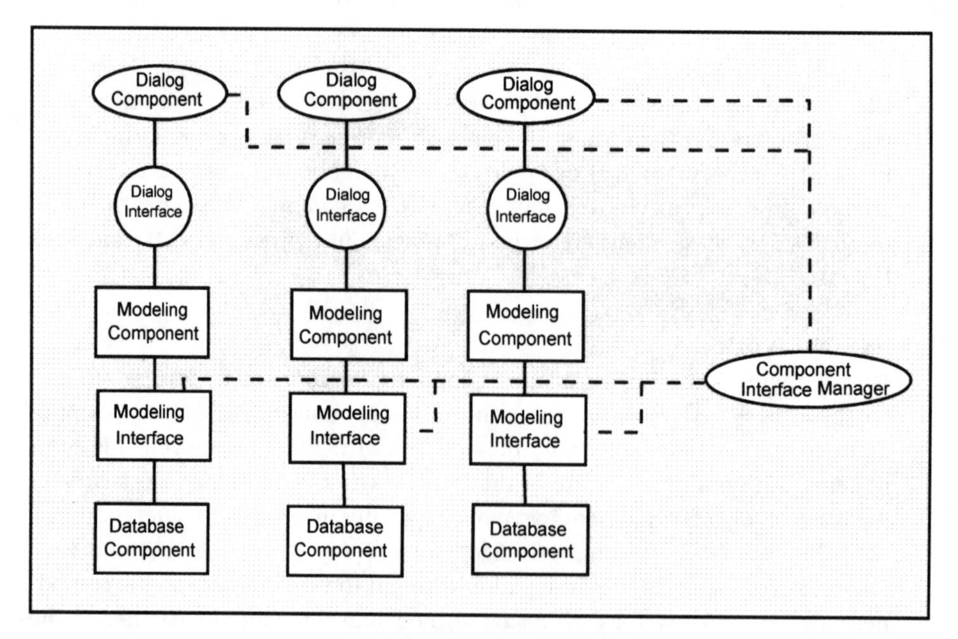

Figure 4.4 DSS-Network Architecture (Based on Sprague and Carlson, 1982)

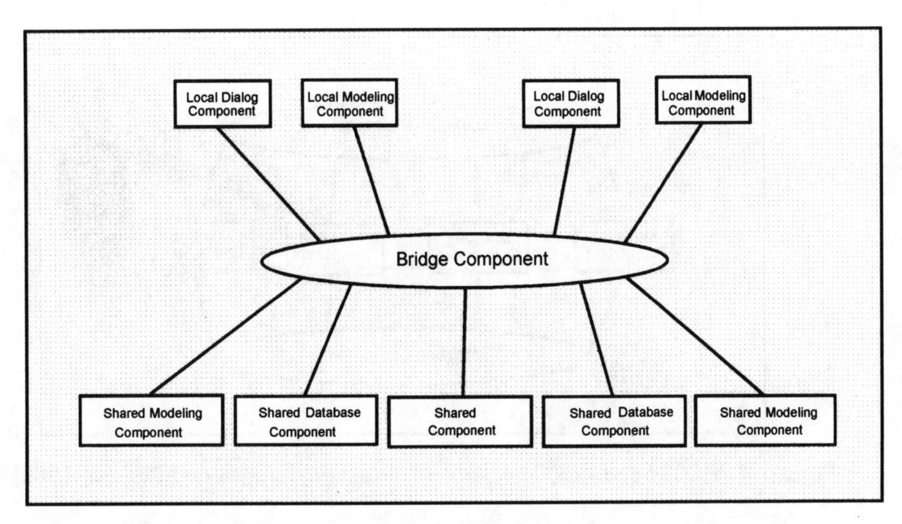

Figure 4.5 DSS-Bridge Configuration (Based on Sprague and Carlson, 1982)

The DSS-Sandwich

This configuration strategy integrates the dialog, model-base, and database elements of the DSS using a single dialog and database component connected to a range of modeling components. In this design, each modeling component included in the DSS shares the same database and uses the same dialog interface. According to this configuration, communication between data and models is handled by the shared database component. Control information is communicated to modeling components through the shared dialog generator. Although standard interfaces are provided by using a single dialog and database component, each modeling element has to be developed or modified to fit into the two preexisting interface styles. The DSS-Sandwich is illustrated in Figure 4.6.

The DSS-Tower

This style of DSS configuration offers component modularity and flexibility that can support a range of hardware devices and source databases while still maintaining a simple interface format among the three functional elements of the DSS. The tower architecture is designed primarily for a single operating environment at any level on the tower, with only a single dialog and database component present. According to this design, dialog, modeling, and database components are layered rather than intermixed using an interfacing strategy similar to that found in the sandwich architecture. Unlike the sandwich design, however, the tower architecture separates the dialog and database component into two parts. The basic structure of the DSS-Tower architecture in shown in Figure 4.7.

The common thread connecting each of the four architectures reviewed above is that in each, the database component plays a central role in interfac-

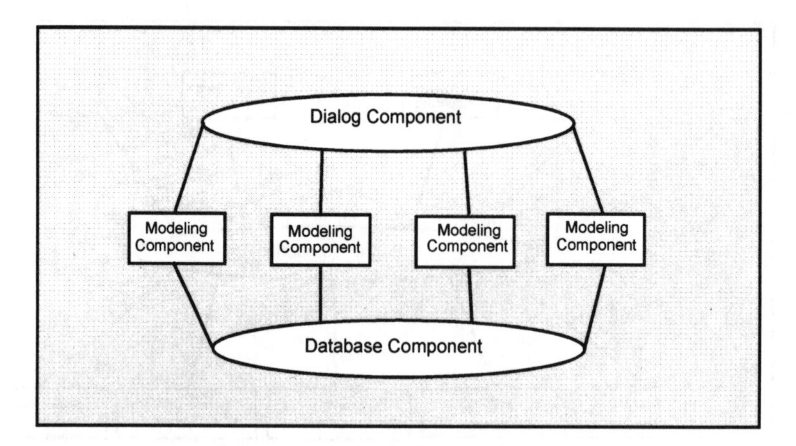

Figure 4.6 DSS-Sandwich Configuration (Based on Sprague and Carlson, 1982)

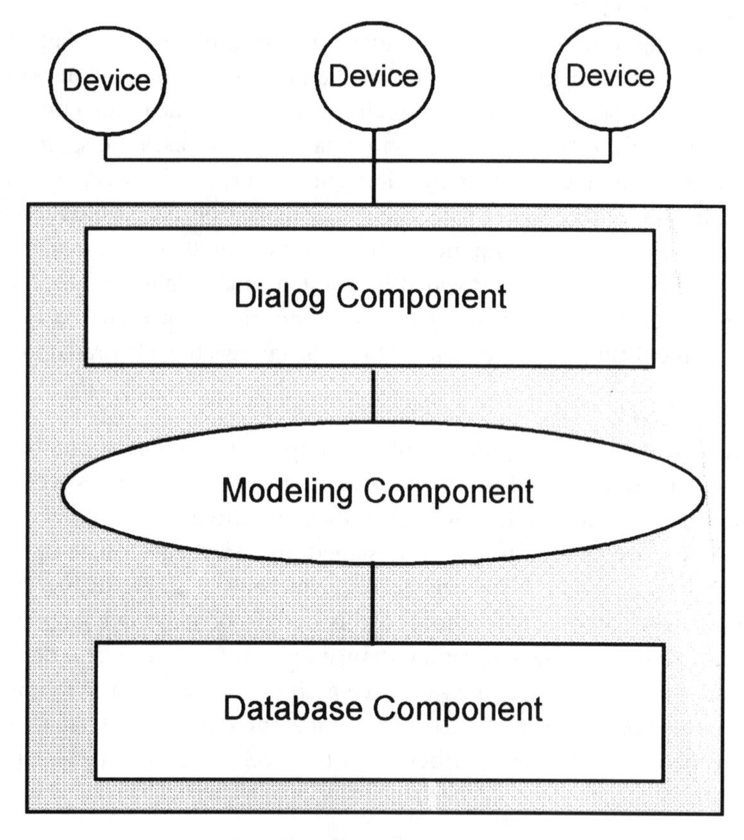

Figure 4.7 DSS-Tower Configuration (Based on Sprague and Carlson, 1982)

ing dialog, modeling, and data in the DSS. The use of the database component in this capacity has three support advantages (Sprague and Carlson, 1982):

1. by separating out parameters of models and dialog, the two components are easier to modify,
2. by storing the parameters in the database, database management functions are provided that do not have to be duplicated in other components, and
3. by storing parameters as data, sharing among components is greatly simplified.

Supporting Environmental Decisions

The concept of an ill- (or semi-) structured problem is the key idea connecting decision support systems to the tasks related to environmental decision making. Because there are many situations encountered in environmental planning and management where there is no known or clear method for solving a problem, or where the nature of the problem itself is often complex and uncertain, adopting decision approaches that can accommodate lack of structure offers the decision maker the unique opportunity to test alternatives and compare alternative selection methodologies. Examples of this include decision problems related to land evaluation, developmental suitability analysis, risk assessment, and environmental impact assessment. In each of these instances, the human analyst/decision maker investigates and the computer supports key tasks by furnishing pertinent information. This unique interaction creates a human-computer decision-making system that provides a level of speed, flexibility, and analytical power that a decision maker working in isolation cannot possess. A decision support system, by facilitating direct interaction with data and models, heightens the preferences, judgments, intuition, and experience of the decision maker in the search for a solution. Therefore, through the DSS, information search, problem definition, computation, and data manipulation are enhanced, which enables the problem to evolve rapidly.

To illustrate this point, consider the decision problem of siting a complex land use and evaluating its environmental impact. In order to determine the appropriateness of a site, the decision maker may want to know how the activity will affect air and water quality, vegetation and wildlife habitat, and natural resources, and whether it will create potential land use conflicts or induce growth. Determining the appropriateness of a given location for the activity in question would require linking data related to the candidate sites with models that explain or simulate the various environmental processes or features involved in the decision. For example, with respect to air quality, the following three simple questions must be addressed and can only be resolved through a modeling exercise.

1. Will the activity result in emissions in the atmosphere and, if so, at what level or concentration?
2. Will the activity contribute to a degradation in air quality above emission standards?
3. Will the activity cause changes in the chemical and physical composition of the atmosphere?

Adding other environmental considerations expands the number of questions that must be answered and moves the information requirements of the decision process squarely into the science of "what if." For instance, to consider socioeconomic effects, questions concerning land use disruptions, changes in the economic base of the region, traffic flow, population densities, and employment must be addressed before an environmentally sound decision can be made. Answering the questions draws heavily upon the decision maker's ability to forecast changes induced by the action. Such forecasting relies on the application of various descriptive and predictive models; fingertip access to critical environmental, economic, and population data; and a means to organize it all into a form that can be applied in a specific analytical methodology: a decision support methodology.

The potential of decision support systems has been explored in a number of contrasting environmental applications (Labadie, 1986; Crosslin, 1991; Connor and Allen, 1994). These examples have culminated in the implementation of computer-based systems designed to address environmental problems (Wood et al., 1988; van Herwijnen et al., 1993; Brown et al., 1994; McClean et al., 1995) as well as in studies that attempt to validate the appropriateness of environmental DSS (Sequeira et al., 1996). In each case, the systems identified specifically link simulation modeling and decision analysis methods with the database component of a geographic information system (GIS). This specific aspect of a DSS points to an important extension of the DSS paradigm and introduces the spatial component as an active feature of a decision support methodology.

Given the inherently geographic nature of the environmental decision-making problem, the integration of DSS and GIS broadens the applicability of both technologies. The issues related to linking DSS and GIS as a single automated support technology have been examined by Fedra and Reitsma (1990). As their paper illustrates, there are several ways to accomplish GIS-to-DSS integration depending on the degree to which the decision problem encompasses a spatial dimension. Integration can be achieved at different levels. To illustrate this point, consider the example of information retrieval and display. When focused on these two operational data needs, the GIS functions to supply data to the model-based component of the DSS and then receives the output generated by that component for the purpose of display or to produce a hard copy printout. However, applying the GIS in this capacity leaves its analytic capabilities unused. This reduces the GIS to the primary role of a database management system and a data/map visualization tool.

The possibilities of true spatial decision support and the opportunities for strengthening the connection between geographic information analysis and decision making have been active areas of research for nearly a decade (Gould, 1989; Densham and Goodchild, 1989; Walsh, 1993; Srinivasan and Engel, 1994). In a review exploring the concept of spatial decision support, Densham (1991) defines this information technology as a special class of DSS designed specifically to support a decision process involving complex spatial problems. In this context, a spatial decision support system (SDSS) facilitates the analysis of geographical information with specific capabilities for handling and modeling spatial data and spatial processes that separate this branch of decision support from the typical DSS. These unique aspects of a spatial decision support system include capabilities to accommodate:

- the input and organization of spatial data,
- the representation of complex spatial relationships,
- specialized analytical techniques for spatial analysis, and
- specialized methods for outputting spatial data in a range of formats, including maps.

These four characteristics give the SDSS an iterative, integrative, and participative functionality. However, Densham (1991) is quick to point out that spatial decision support systems, because of their design complexity and specialization, are typically implemented for a narrow, limited problem domain. The specialized nature of SDSS is well illustrated by Armstrong et al. (1991). The SDSS described in this paper was designed to address a clearly defined facilities location problem with all system functionality incorporated in its database, model base, and interface devoted to supporting this specific type of facilities location problem. In this example of SDSS, the focus of system development concentrated on the problem of determining the appropriate number and location of centralized service facilities (that is, Area Education Agencies) to support a geographically dispersed population. The system developed for this application had four interrelated components:

1. **the database**—serving as the foundation of the SDSS,
2. **a set of data transformation utilities**—designed to preprocess data for analysis,
3. **analytical and modeling algorithms**—explaining specific shortest-path and location-allocation modeling software designed to derive, evaluate, and test candidate facility locations, and
4. **mapping and report generation routines**—included to prepare the resulting information obtained in the analysis as output.

With this system so described, decision makers were able to explore the facilities location and population/demand allocation problem by examining the results of analyses using different parameters as decision criteria. In the example given by Armstrong et al. (1991), with this specialized type of spatial decision support decision makers are provided with an experimental setting where:

- the number of candidate sites could be varied,
- the maximum travel distance constraints could be adjusted,
- the minimum population (demand) threshold levels could be modified, and
- other conditions specific to the problem could be changed in an experimental fashion.

In this application of SDSS, the generate and test qualities afforded by the system provide decision makers with a greater opportunity to reach a consensus on a solution. By employing the system to create a set of possible solutions, the relative merits of each can be evaluated and the "optimal" location, whether expressed from a political, economic, environmental, or social perspective, can be determined. This feature of SDSS methodology can be applied to a wide range of spatial problems that open up new opportunities for the design and implementation of GIS-based decision support.

GIS-Based Decision Support

The widespread adoption of GIS technology suggests that despite its analytical limitations, geographic information systems offer decision makers a standard platform for the storage, retrieval, and analytic manipulation of spatially referenced data. In this capacity, GIS technology has generally been successful when applied to problems that characterize clearly defined questions and measurable outcomes. It is not surprising, therefore, that expanding GIS technology to support a broader spectrum of decision problems is an active research frontier. Since GIS is presently employed at some level of decision support, integrating improved methods of spatial analysis and dynamic modeling can only enhance the possibilities as well as the potential for GIS-based decision support.

Connecting dynamic modeling environments with GIS has been pursued from several different directions (Grossman and Eberhardt, 1992; Hazelton, Leahy, and Williamson, 1992). In general, bringing dynamic modeling to GIS is viewed as a necessity in order to: (1) take GIS beyond its present status as a data retrieval and sorting engine, (2) enable GIS to function as a decision support tool, (3) provide GIS with other models of reality, and (4) create a more unified platform for environmental decision making. However, as Grossman and Eberhardt (1992) note, the connection of generic models to GIS is methodologically difficult because the "bookkeeping" is complex. Conflicting data formats, contrasting schemes of data representation, and the need for data transformation create a complex web of functional support that makes the notion of "plug-in" modeling a challenge. This fact has presently led to the conclusion that "seamless" integration between GIS and modeling systems is impractical due largely to the incompatibilities surrounding the conceptual frameworks of the two systems. This fact, however, should not preclude other approaches.

The most common method of linking GIS with dynamic models is system interfacing. With system interfacing, rather than being fully integrated, the GIS and modeling environments are connected via a file exchange interface. The interface serves as a flexible conduit that permits data to be passed between systems and reduces data incompatibility problems. Three leaves of GIS-to-model-based integration have been suggested (Tim and Jolly, 1994):

1. ad hoc integration,
2. partial integration, and
3. complete integration.

Ad hoc interfacing is a form of GIS-model integration where the GIS and the model have been developed separately. This type of interfacing is common in GIS applications that utilize simulation models developed as stand-alone software systems. Input data required by the model is extracted from the GIS database, allowing the model to run independently. This loose form of integration places little demand on either the GIS or the model; however, the potential for error is high since there is no mechanism to control the quality of data leaving the GIS or to insure the integrity of model output when passed back to the GIS.

The method of partial integration is an interfacing scheme that can take two slightly different forms. One form of partial integration requires development of a GIS database in order to provide specific inputs to an existing model. The second method of partial integration requires the model to be built on top of an existing GIS. With this form of interfacing, the model input must be structured according to the method used to store data in the GIS database. In either example of partial integration, the GIS supplies input data for modeling applications and accepts model results that can be subject to further GIS processing or configured for output using GIS display and mapping facilities.

The final method of model-to-GIS integration brings the model directly into the software design of the GIS. This form of integration describes the complete merger of the two technologies and suggests that the GIS and the model have been developed in close interaction within a single operating environment. Examples of these approaches to system interfacing are described in detail by Tim and Jolly (1994).

The Knowledge Connection

The application of data and models to the solution of an ill- or semi-structured problem, while critical to the concept of decision support, fails to incorporate adequately the expertise of the decision maker. Expert judgment, expert opinion, and the subjectivity they introduce are inherent features of the environmental decision-making process and are deeply embedded in nearly every facet of decision making. Although decision support technology helps the expert decision maker by providing access to data and models that permit subjective judgment and technical knowledge to blend and interact with the machine environment, the expertise is neither explicit to the system nor formalized in any comparable way. Therefore, while a DSS or SDSS contains "knowledge" in the form of data, text, variables, models, reports, and graphics, it lacks expertise and cannot aid the decision maker who may be ignorant of one or more aspects of the problem. Thus, information technology needs the capacity to store and preserve the specialized knowledge of the human expert and introduce judgment and reasoning to the information system and decision support environments. With such a capacity, an information technology would be rendered "intelligent," which would expand the range of problems it could address. The groundwork to realize this intelligent transformation is presented in Chapter 5.

5 Applied Artificial Intelligence

Over the past decade, artificial intelligence has moved out of the laboratory and into mainstream science, industry, and engineering. Because of its potential to make machines function more intelligently, artificial intelligence is very appealing. While the prospects for "true" machine intelligence are still forthcoming, a computer today can perform limited tasks that model intelligent behavior, thus providing the opportunity for the user of that computer to function more intelligently as well. Recent developments in the commercial application of artificial intelligence have produced a promising number of technologies that offer the potential for the user of a computer to function more intelligently, and these have important implications for environmental decision making. This chapter reviews the fundamentals of artificial intelligence and examines the contributions of applied artificial intelligence research directed towards the problem of environmental decision making. This discussion introduces the expert system, exploring its development and application as a decision tool and as a complement to existing information technologies such as GIS and DSS.

Data, Information, and Knowledge

When considering information technology and its role as a decision-making aid, emphasis is usually given to techniques designed to facilitate the storage, retrieval, and management of data. In the areas of environmental analysis and management, this focus is clearly illustrated by the steady interest in GIS and other database systems. Yet, there is a growing awareness that merely processing data is not enough unless that data can be organized in such a way that it becomes meaningful to the needs of the decision maker. When this happens, data is transformed into information and the technologies developed to expedite its delivery enhance simple data processing by adding capabilities that permit analysis and modeling with that data. Such coupling of advanced processing and simulation methods transforms the data system into an information system that is more adept at meeting the information-specific demands of the decision maker.

Sometimes situations emerge where data and information are, by themselves, insufficient considering the complexities of the decision-making process. In these instances, data may be available and may be adequately recast into useful information, but the decision maker lacks the knowledge or understanding to interpret that information or place it in its proper context. When circumstances such as this arise, there is a demand for knowledge and there is a need to access that knowledge in much the same way that data or information is accessed in a machine environment. Unlike data, which can be described using raw numeric or alphanumeric strings, knowledge is much more abstract and difficult to define. Indeed, knowledge can imply any or all of the following traits:

- understanding,
- learning,
- perception,
- expression,
- skill,
- cognition/recognition, and
- organized information.

Similarly, knowledge, as it relates to environmental decision making, also suggests a level of expertise that characterizes an individual engaged in the process of solving problems.

From a machine perspective, knowledge, like data and information, can be organized, stored, accessed, and manipulated as if it were a physical quantity. And knowledge encoded into the computer is significant for two reasons. First, in a very simple way, it makes the computer a repository of domain-specific knowledge. Knowledge can therefore be organized into a logical structure that enables the machine to process it, suggesting that in a limited way a computer can use knowledge and function in an entirely different way and address very different types of problems. Secondly, encoded knowledge is a resource that can be made available to the decision maker. Knowledge, when expressed in the form of a computer, is a tool that makes the decision maker function differently as well. With access to knowledge, the decision maker becomes more knowledgeable and capable of making better decisions.

Numerous disciplines lend expertise and influence environmental decision making. For example, environmental problems can include ecological, economic, and policy considerations, and it is functionally impossible for a given decision maker to have expertise in all of these areas. Add to this the growing realization that the majority of environmental professionals possess very specialized knowledge and the end product may be narrowly trained decision makers incapable of navigating all aspects of a problem. As a result, critical environmental factors are too often ignored or overlooked, revealing large gaps in knowledge that erode the effectiveness of environmental decisions. Building this knowledge into the decision support apparatus can reduce that risk of "ignorance," extend the decision maker's knowledge, and thereby improve the decision-making process. Thus, artificial intelligence has tremendous potential power.

Essential Artificial Intelligence Concepts

An all-encompassing yet precise explanation of artificial intelligence remains elusive. Primarily, however, the field of AI is concerned with designing and programming machines to accomplish tasks that human beings accomplish using their intelligence (Schutzer, 1987), or, as Rich (1983) suggests, AI is the study of how to make computers do things that, for the moment, people do better. From an entirely applied perspective (Winston, 1984; Winston and Prendergast, 1984), the three major goals of AI are:

- to make computers more useful,
- to understand the principles that make intelligence possible, and
- to make machines smarter.

Yet, to the environmental decision maker, the most important characteristic of AI is that it represents a problem-solving paradigm that differs from the more traditional algorithmic approaches. Problems addressed by AI are solvable by human beings but do not lend themselves to structured computation. Problem solving using AI relies on symbolic processing, heuristics, inferencing, and pattern matching, as opposed to the quantitative and numerical representation schemes used in traditional computer programming.

A machine exhibits "intelligence" through the mechanisms that perform symbolic processing, inferencing, and pattern matching, and by our applying heuristics to control these procedures. Exactly what constitutes machine intelligence is difficult to state in absolute terms. Therefore, it is more practical to establish criteria that can serve as evidence of the presence or absence of intelligence (Fetzer, 1990). When viewed in this manner, intelligence becomes a behavior that a machine can exhibit, such as one or more of the following abilities (after Turban, 1992):

- to learn or understand from experience,
- to interpret ambiguous or contradictory information,
- to respond quickly and successfully to a new situation,
- to use reason in solving problems,
- to deal with perplexing situations,

- to understand and infer in ordinary and rational ways,
- to apply knowledge to manipulate the environment,
- to acquire and apply knowledge,
- to reason, and
- to recognize the relative importance of different elements in a situation.

However, knowledge is required for a machine to exhibit intelligent behavior (Rich, 1983). Although knowledge is indispensable to any problem-solving application, it is also voluminous, difficult to accurately characterize, and constantly changing. AI techniques exploit knowledge but also require that it be represented in such a way that (Rich, 1983):

- it captures generalizations,
- it can be understood by people who must provide it,
- it can be easily modified to correct errors and reflect changes,
- it can be used in many situations even if it is incomplete or not totally accurate, and
- it can be used, by virtue of its size, to assist with the narrowing of possibilities and the evaluation of options.

Human problem solving is also knowledge-dependent. In reviews outlining the characteristics of human problem-solving and expertise, Anderson (1985) and Johnson (1983) identify three important categories of knowledge that align AI more directly with the behavioral qualities of decision making. This categorization of knowledge underscores the need to develop ways to employ and preserve specialized "problem-specific" knowledge for application in machine environments. The three categories of knowledge are summarized below.

1. **Domain Knowledge**
   Domain knowledge is the body of operative knowledge underlying the behavior of experts (Klein and Methlie, 1990). This knowledge is specific to the knowledge structure of the domain, and through training and experience, these structures are compiled into complex schemas and representations.
2. **Unconscious Knowledge**
   As decision makers master domain knowledge in order to perform a task efficiently and accurately, they lose awareness of what they know. Thus, experts lose the ability to explicate their full knowledge, which suggests that certain aspects of learning are not always available to conscious awareness.
3. **Theoretical and Experiential Knowledge**
   Human problem solving and expertise also draw on the conceptual, analytical, and experiential knowledge acquired through training and practice. Experience and knowledge of theory are unique qualities of an expert and show a depth of understanding in the problem domain.

Producing computer programs that simulate or employ human problem-solving strategies depends on the degree to which each of the three categories identified above can be acquired and represented. Because an AI system is capable of not only storing and manipulating data, but also acquiring, representing, and manipulating knowledge, operating with intangibles such as ideas, concepts, and their relationships depends on a set of formalisms that can represent and manipulate abstract entities (Schutzer, 1987). In this sense, manipulation includes the ability to deduce or infer new knowledge and new relationships from existing knowledge. It also implies the ability to use representation and manipulation skills to solve complex problems that are often non-quantitative in nature. In knowledge-rich but algorithmically poor disciplines such as environmental planning and management, alternative problem-solving strategies may hold promise.

Given the observation that environmental decision making frequently involves complex interactions among factors that are: (1) often known only to the decision maker, (2) qualitatively defined, and (3) imprecise or uncertain in their behavior, developing techniques that can draw on and incorporate knowledge not only renders the information technology smarter and more useful, but also produces an information resource, expressed in the form of knowledge, that decision makers can access and employ when their own domain knowledge is insufficient to the task. Tapping into this knowledge resource with a focus on decision making requires a more detailed explanation of four key AI concepts introduced in the discussion above.

1. **Symbolic Processing**
   The notion of a symbol is so pervasive in the current theory and practice of AI that its importance is easily overlooked. However, the notion of symbolic processing forms the critical link between AI applications and formal systems of logic and mathematics (Jackson, 1990). A symbol is something that stands in place of something else. That something else may be an object or a concept, but the symbol itself is always a physical entity. When human experts solve problems, particularly the type considered appropriate for AI, they do not typically solve sets of equations or perform other laborious mathematical computations (Turban, 1992). Rather, they select symbols to represent the problem concepts and apply various strategies and rules to manipulate these concepts. The symbols can be combined to express meaningful relationships and are routinely manipulated by a computer. Symbolic processing, therefore, is concerned with symbolic, nonalgorithmic methods of problem solving.
2. **Heuristics**
   A heuristic is an informal type of knowledge commonly thought of as a "rule-of-thumb." Decision makers frequently apply heuristics when processing information as approximations of the solution, particularly in sit-

uations where perfect data or perfect answers are not required or available. By using a heuristic, the decision maker need not rethink a procedure completely every time a similar problem is encountered. This saves time and effort, and when applied in a machine environment, heuristic reasoning can process information, obtaining solutions with varying degrees of certainty. Although the solutions might not be optimal, they may be considered good enough for a process that may be iteratively refined.

3. **Inferencing**
   To infer means simply to derive a conclusion from a set of facts or premises. Inferencing is that process and explains a type of reasoning from facts and rules that uses heuristics and other methods of search. Through inferencing, a machine approximates human reasoning behavior based on a specific search paradigm that facilitates the comparison of symbols used to represent knowledge.
4. **Pattern Matching**
   As a type of inferencing strategy, pattern matching attempts to describe objects, events, and processes in relation to their qualitative features and logical or computational associations. An object, fact, or event is characterized by a set of variables/symbols that define it as a pattern. By comparing patterns, the computer performs a search. Pattern matching may also be used to invoke rules or heuristics to better direct the search.

The four root AI concepts defined above illustrate the requirements that an AI program shares, and that further distinguish it from a conventional computer algorithm, namely;

- a knowledge representation framework, and
- a problem-solving and inferencing strategy.

Based on these two defining conditions, the applications best suited to AI technology are problems where:

1. no well-defined algorithmic solution exists,
2. some decision-making information is available,
3. uncertainty exists to a large degree, and
4. some information is missing or incorrect.

The Applied Value of AI

As with any information technology, the practical application of AI must be justified to some degree. The environmental decision-making arena is no exception. Up to this point in the book, decision makers have been presented with an array of technologies, all proposing to help them make better decisions. Introducing AI as another method of support may cause you to ask, Why AI? While the answer to this question may seem self-aggrandizing, it is not trivial. If human intelligence is contrasted with artificial intelligence, the value of AI to the decision maker can be better understood (Turban, 1992). Generally, AI is recognized as having several applied advantages (Kaplan, 1984) including the following:

- it is more permanent—human expertise is transient (that is, people move, forget, and so forth),
- it offers ease of duplication and dissemination,
- it can be less expensive,
- it is consistent and thorough, and
- it can be documented.

Conversely, there are several areas where "natural" (human) intelligence enjoys the comparative advantage, including:

- creativity,
- use of sensory experience, and
- the context of human experience.

Although the general rationale presented here to justify the adoption of AI technology may suggest its potential, its role as an environmental problem-solving tool requires a more careful examination.

When exploring the field of AI, one quickly realizes that AI is a science that involves many diverse subject areas including cognitive psychology, computer science, engineering, philosophy, and linguistics. Through the combined efforts of researchers in these disciplines several important AI technologies have emerged (Table 5.1). Of the technologies listed in Table 5.1, one in particular, the expert system, has made a significant contribution to environmental analysis and decision making. The fundamentals of expert system technology and its application to environmental decision making will be explored in the following sections.

Table 5.1 Artificial Intelligence Technologies

Natural Language Processing
Intelligent Tutor
Computer Vision
Expert Systems
Automatic Programming
Speech Understanding
Robotics
Machine Learning
Game Playing

Expert Systems

A detailed literature examining the design and development of expert systems has evolved rapidly over the past decade, as have general works exploring the applications of this technology in environmental management and natural resource planning (Hushon, 1990; Kim et al., 1990; Wright et al., 1993). Essentially, an expert system is a computer-based technology that emulates the decision-making ability of a human expert. Building on this basic definition, an expert system is more precisely defined as a system that employs human knowledge captured in the form of a specialized computer program and uses that knowledge to solve problems in a clearly defined domain. As such, an expert system imitates the reasoning processes of a human expert and obtains a solution similar to that which would be arrived at by the human expert. In this regard, an expert system possesses two important characteristics (Weiss and Kulikowski, 1984):

1. it handles real-world, complex problems that typically require an expert's interpretation, and
2. it solves these problems using a computer model of expert human reasoning.

Given these two defining characteristics, an expert system captures enough of the human specialists' knowledge so that it will be capable of solving problems expertly. From a more functional perspective, an expert system performs tasks typically undertaken by a human expert and addresses problems for which clear algorithmic solutions do not exist (Parsaye and Chignell, 1988). Therefore, the power of an expert system is not necessarily the search paradigm or the specific reasoning method used in programming its structure. Rather, its power is derived from the knowledge it embodies. Formalizing the knowledge of a human expert and codifying it into an expert system is a critical aspect of expert system development that involves extracting the "rules-of-thumb" and applied knowledge gained by the expert through training and experience. When considered relative to the knowledge it contains, the expert system represents a knowledge-based program that facilitates a machine's emulation of the reasoning and decision-making behavior of humans.

As a problem-solving technology, the success of an expert system application is related to the type of problem it is designed to address. In general, an expert system displays less breadth of scope than a human expert and describes a narrower, more problem-specific focus. Initially, this may appear to be a limitation, but a narrow focus implies specialization and competence, which can insure that the system achieves a satisfactory level of performance in its domain. This is important, since the knowledge contained in an expert system may originate from a human expert, books, research papers, or other documents. Hence, any limitations of the system are imposed by the problem domain and not the technology. The problem domain of the expert system is generally thought to reflect the special problem area in which an expert can solve problems, and, like the human expert, the expert system is designed to perform in only one problem domain (Giarratano and Riley, 1989).

Expert System Structure

Just as a human expert applies knowledge and reasoning to arrive at a conclusion, an expert system relies on knowledge and performs reasoning. The type of reasoning displayed by an expert system simulates a human expert's reasoning by combining pieces of knowledge in a specific way. Precisely how this knowledge is combined depends on the structure of the expert system. To help you grasp the issues surrounding expert system structure and the design elements that form an expert system, here is a brief and highly simplified description of a human expert, defined in component form.

A human expert can be defined as a collection of components, listed below, that provide a level of functionality that facilitates problem-solving:

- memory,
- reasoning center, and
- language center.

Memory is where the facts, structures, and rules that constitute expert knowledge reside. The reasoning center represents the cognitive processes of logic, deduction, and inference, which constitute thought and reasoning. The language center directs communication and enables the expert to interact with other information sources. As a problem is presented to the human expert through the language center, it is interpreted, and the relevant facts and knowledge are accessed from memory and applied in an exercise that draws on reasoning and thought. For a machine to exhibit these qualities exclusive of expert performance, it must be given analogous components. In an expert system, these fundamental elements are the knowledge base, inference engine, user interface, and explanation facility. Together, these four components form the nucleus of an expert system (Figure 5.1). In the following sections, each of these elements will be reviewed in detail, beginning with the knowledge base.

The Knowledge-Base Component

The knowledge base of an expert system is similar in concept to the database component of an information system. The knowledge base contains the facts and rules that define expertise in the system's problem domain. A typical knowledge base includes two fundamental expressions of knowledge: (1) facts, which characterize the formal theories and specific relationships of the problem area in which the system is designed to perform, and (2) rules, which describe heuris-

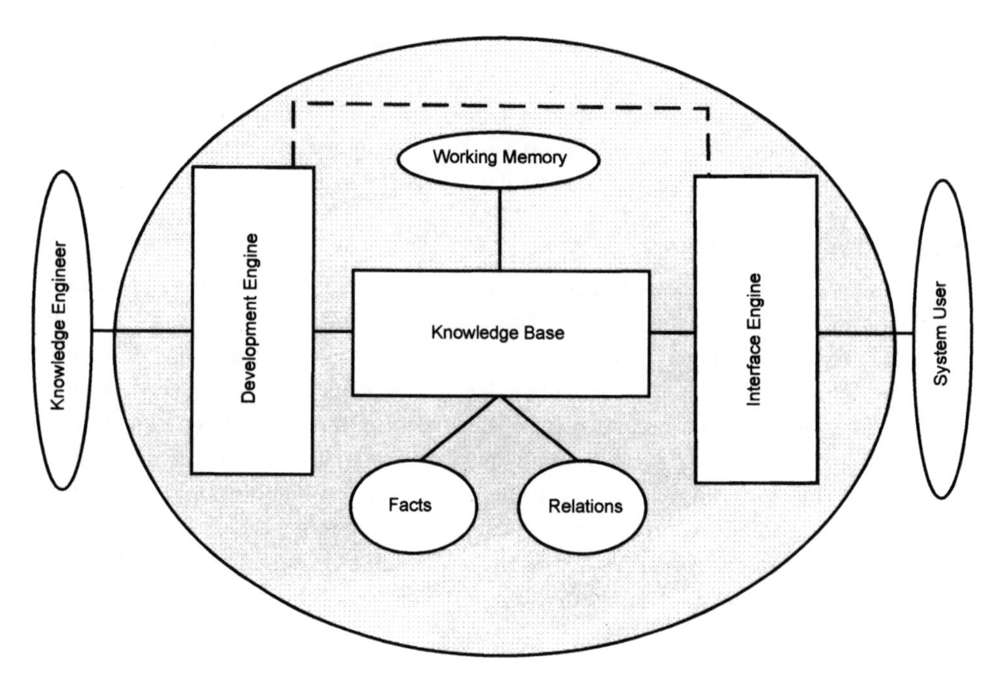

Figure 5.1 Basic Components of a Knowledge-Based Expert System

tics that direct the use of knowledge and express informal human judgment in the problem domain. Knowledge in an expert system can be structured in several different ways. Two of the more common ways include rule-based methods and frame-based methods (Waterman, 1986).

Rule-Based Methods

The rule-based method of knowledge representation expresses knowledge in the form of IF-THEN action statements. These IF-THEN clauses, referred to as production rules, take the following general form:

IF [a condition or qualification is **TRUE**]
THEN [a specific action is taken],

or the more simplified version:

If [premise] **THEN** [conclusion].

Given this representation scheme, a rule is proven true when a premise of the rule matches known facts. This match confirms the conclusion and invokes the clause contained within the conclusion as fact. For example, in an expert system developed to assist the land evaluation process, a rule may exist that is designed to determine soil limitations for single-family dwelling units using septic wastewater treatment systems. Such a rule might take the form:

IF proposed use is single-family dwelling

AND septic system is yes

AND permeability is 2.0 to 0.2 inches per hour,

THEN soil limitations are moderate.

Similarly, a rule to evaluate soil erosion potential might express knowledge as:

IF slope ≥ 25 percent

AND soil texture is medium-fine-grained

AND land cover is barren,

THEN erosion potential is severe.

In either example, when the current problem situation satisfies or matches the IF clause of the rule, the action/conclusion specified by the THEN clause of the rule is performed. In a rule-based expert system, the action or conclusion may invoke several system behaviors such as (1) triggering an event external to the system (for example, causing text to be printed at the terminal), (2) directing program control (for instance, causing a particular set of rules to be tested), or (3) instructing the system to reach a conclusion (for instance, adding a new fact or hypothesis to the database) (Waterman, 1986).

As a means of knowledge representation, rules provide a fairly straightforward way of organizing all of the concepts, facts, and related information specific to a problem into a logical structure that is easily understood by both the developers and users of the expert system. Rules also provide a natural way of expressing inferential knowledge, and describe a convenient method for describing processes driven by complex and rapidly changing environments. In addition, rule structures are very adaptable. When structured as a sequence of rules, knowledge offers nearly unlimited flexibility, which permits the nesting of several conditions or facts in a single rule (as illustrated in the examples given above) in order to reach complex conclusions or perform detailed actions. This feature is important to the user of the expert system because the number of rules forming the knowledge base, together with their complexity and nesting, helps define the system's depth. Because there is no theoretical limit to the number of conditions that can be evaluated in a single IF-THEN clause, a series of very complex relationships can be tested before an action is taken or a conclusion is assigned.

Frame-Based Methods

Frame-based methods of knowledge representation use a network of nodes connected by relations and organized into a hierarchy. While rules are an effective means of representing inferential knowledge, they are not well suited to the representation of structural or descriptive knowledge (Klein and Methlie, 1990). A frame, according to this schema, is an object-oriented view that allows descriptive knowledge to be partitioned into discrete structures, each possessing individual properties of some larger whole. In general, frames can be used to represent broad concepts, classes of objects, individual instances of objects, or parts of objects. In addition, specific information may be attached to each frame and may include instructions about how to use the frame, what should follow next, or how to handle conflict. An excellent discussion of frame-based methods of knowledge representation can be found in Parsaye and Chignell (1988).

Essentially, frames form useful packets of knowledge in which specific attributes of objects, events, or facts are stored in slots. Using a parent-child network relationship, descriptions and representations are formed. These relationships are generally based on inheritance principles, which connect slots in frames to complete the larger hierarchy, which then fulfills a specific characteristic or description of a feature.

Thus, frames are the building blocks of an object system that represent the generic characteristics of items that comprise the domain. For example, in a land use system, a frame structure might be used to describe residential, commercial, and industrial parcels with attributes stored in slots that hold information pertaining to ownership, assessed value, and square footage. In this example, the frame represents the static description of the domain. To illustrate this property of frame-based methods, consider the land use example given above. A hierarchy of frames describing land uses could follow the general structure shown in Figure 5.2. According to this structure, slots in a land use frame would characterize selected attributes similar to those presented in the illustration. However, before a frame can be used, it must be identified as applicable to the current situation. Typically, this is accomplished by matching the frame system with a set of facts. According to this logic, the selected frame is the frame with the greatest number of lower-level slots filled in. From here, the attempt is made to fill in higher-level slots, and if the search fails, another frame is selected.

Whether expressed in the form of rules or frames, the knowledge base embodies the expertise that will be applied by the expert system as it reasons in its problem domain. Selecting and applying knowledge from the knowledge base in order to solve problems is the principal function of the second major component of an expert system: the inference engine.

The Inference Engine

The inference engine has frequently been described as the "brain" of an expert system (Turban, 1992). The inference engine of an expert system is a computer program that contains rules governing a specific methodology for reasoning with the facts and relations contained in the knowledge base, and in the working memory, and for formulating conclusions. In a sense, the inference engine rep-

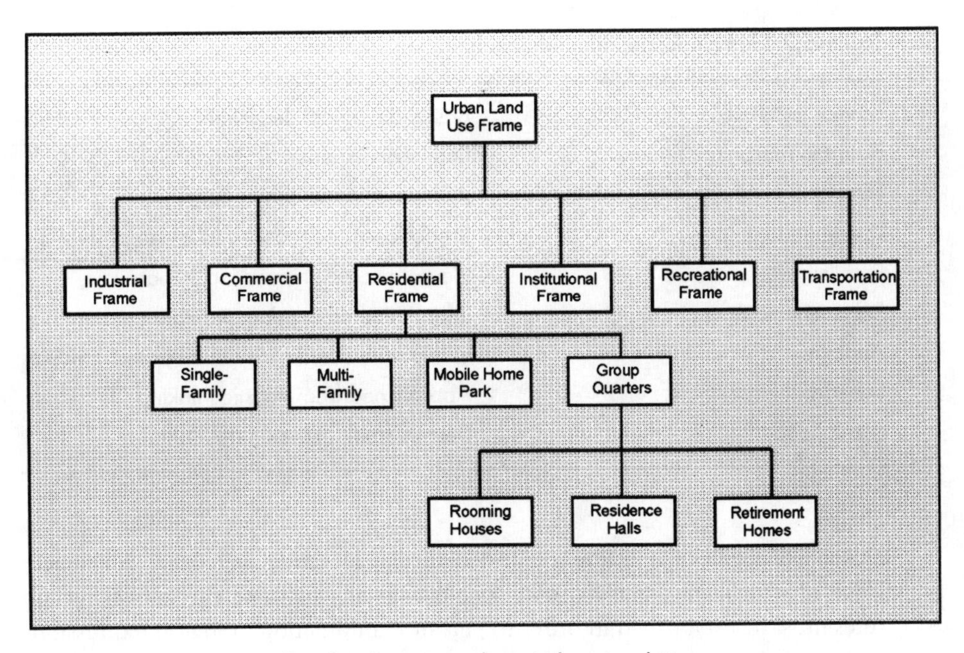

Figure 5.2 Hierarchy of Frames Describing Urban Land Use

resents a store of general problem-solving knowledge that interprets rules and controls the execution of the knowledge base. The knowledge contained in the inference engine implements a reasoning strategy. This strategy provides directions concerning how to apply the system's knowledge through an agenda that organizes and controls the procedure followed in solving a problem during consultations with the system.

Every inference engine contains three fundamental subcomponents:

1. **an interpreter**—that decides how to apply the rules in the knowledge base and how to infer new knowledge,
2. **a scheduler**—that controls the agenda by deciding the order in which the rules should be applied, and
3. **a consistency enforcer**—that attempts to maintain agreement among rules and reduce inconsistencies that can be attributed to conflicting rules.

Through interaction among these components, the strategy that guides inferencing and controls the reasoning process is implemented by the system. Although the concept of reasoning is difficult to define, examining how a human expert reasons and solves problems provides important insight into the development of machine reasoning techniques. Lenat (1982) identified nine modes of human reasoning that influence the major inference and control strategies developed for AI applications (Table 5.2). From these forms of human

Table 5.2 Human Reasoning Methods

Formal methods	Based on logical deduction
Heuristic reasoning	Draws on experience
Focus	Applies common sense
Divide and conquer	Breaks complexity into smaller subproblems
Parallelism	Neural processors working in parallel
Representation	Information-organizing strategies
Analogy	Works with associations and related concepts
Synergy	Approached from holistic thinking
Serendipity	Derives from fortuitous accidents

reasoning, specific AI inference and control strategies have been developed, some of which are given in Table 5.3. Within any of these categories, certain rules must be established in order to permit manipulation of logical expressions and to generate new expressions in a machine environment. Modus ponens, modus tollens, and the resolution principle are three of the more important rules with respect to the development of an inferencing strategy (Jackson, 1990).

Modus ponens states that

$$\text{IF}\ (a)\ \text{THEN}\ (b)$$

and (a) is known to be true, then it is valid to conclude that (b) is also true. Symbolically, this relation can be written as

$$\left[\text{A and}\left(\text{A B}\right)\right]\quad \text{B},$$

where (A) and (A B) are propositions in the knowledge base. The opposite of this condition is referred to as modus tollens. Here, given the rule IF (A) THEN (B), and (B) is known to be false, then it is valid to conclude that (A) is also false.

The resolution principle, first described by Robinson (1965), is a method of theorem proving that can be applied to clauses in a rule. In designing an inference strategy, the resolution principle is used to discover whether a new fact is valid given a set of logical statements. Thus, as an inferencing device, it provides a logic for determining whether a hypothesis follows from a given set of premises.

When these procedures are brought together, they permit the derivation of new facts from rules and known facts, and they provide a basis or foundation for programming a machine reasoning strategy or style. Two common reasoning styles can be employed in developing the inference engine of an expert

Table 5.3 Principal AI Inferencing and Control Strategies

Deductive Reasoning

A process by which general premises are employed to obtain a specific inference, describing a reasoning process that flows from a general principle to a specific conclusion.

Inductive Reasoning

A process that applies a number of established facts in order to draw a general conclusion that may be difficult to arrive at and easily changed if new facts are introduced.

Analogical Reasoning

A way of deriving an answer based on analogy using a form of internalized learning that permits recognition of previously encountered experiences.

Formal Reasoning

A process that involves syntactic manipulation of data structures to deduce new facts based upon prescribed rules of inference.

Procedural Numeric Reasoning

A process of describing the application of mathematical models or simulation to solve problems.

Generalization and Abstraction

The drawing of inferences based on logical generalizations using semantic representations of knowledge.

Metalevel Reasoning

The application of knowledge pertaining to the importance of facts and their significance.

---

system: backward chaining and forward chaining. In either instance, the inference engine combines the facts and rules of the knowledge base to reach a conclusion, although that conclusion is obtained following a different logical path. Performing inference, therefore, becomes a matter of establishing the truth or falsity of a goal statement (conclusion). For instance, the statement "Erosion potential is severe" may express one goal in an expert system designed to address the problem of land capability. Within the knowledge base, a series of rules should exist that are connected by the inference engine to produce a line of reasoning that can establish the truth value of that goal. By matching the IF portions of the rules to the facts, an inference chain is produced that indicates how the system employed its rules to infer a given conclusion. As noted previously, two methods of reasoning with rules exist that can be used to formulate an inference chain. These methods are reviewed below. However, for a more detailed discussion of inferencing styles, the reader is directed to Payne and McArthur (1990).

Backward Chaining

Backward chaining can be defined as a goal-driven approach for controlling inference. With backward chaining, inferencing begins with an expectation or hypothesis and seeks evidence that supports or refutes that hypothesis. Therefore, if the current goal is to determine the fact in the conclusion of the rule, the inferencing process attempts to determine whether the condition in the rule's premise matches the present situation. According to this strategy, the system takes an initial set of goals, and the rules comprising the knowledge base are invoked in reverse order. Inferencing begins as the system examines a limited number of rules, concentrating exclusively on their conclusions. Inferencing continues as the system conducts a search of each rule's premise to determine which of the goals are satisfied.

Forward Chaining

Forward chaining is a data-driven inferencing strategy. A forward-chaining strategy begins with the available information as it arrives or with an assertion. Using this information, the system then attempts to draw a conclusion. The system examines the problem by looking for the facts that match the premise of its rules. In forward chaining, the system does not start with any particular goal defined, but with a subset of evidence. It proceeds to invoke the rules in a forward direction, continuing its comparative evaluation until no further rules can be invoked.

Working Memory

The working memory of an expert system is an area of memory in the computer reserved for loading the facts and rules of the knowledge base and for implementing an inferencing strategy. Because the knowledge base of a detailed expert system can be large, and because the requirements of complex inference chaining can become extensive, expert systems can make heavy demands on a computer's memory resources.

Explanation Facility

The explanation facility provides a critical link between inferencing and the user's ability to understand how the inferencing process produced a given conclusion. In a manner similar to asking the human expert how or why he or she arrived at an answer, the explanation facility of an expert system supports a justifier or explanation subsystem that enables the system to examine its own reasoning and explain it during consultation. During a consultation session, situations may arise when explanation is needed. In some cases, the system's response, prompts for information, or conclusions may be puzzling or may deviate from what the user expected. Explanation can help the user assess whether the conclusion seems reasonable, and may provide a clue as to the rationale used to draw a particular conclusion. In this way, explanation permits the system to explain a given line of reasoning and to have that reasoning chain examined by the user.

In most knowledge-based systems, explanation can be required either during a consultation or after the consultation has been completed. When called for during consultation, explanation can provide needed clarification when prompting the user for information or when presenting terminology the user may not understand. Here, explanation can define terms, provide a reason why a certain question is asked, and explain why a certain answer is required. Post-consultation explanation gives the system an opportunity to explain how a conclusion was derived, which rules were used to reach that conclusion, why those particular rules were used, and what that conclusion may mean. This form of explanation is important from the user's perspective since it provides some justification for believing the expert system is correct and offers the user the chance to double-check the system's reasoning before using the information gained via consultation.

Because a knowledge-base developer has control over the contents of each rule's reason and each variable's label, the explanation that may be produced by the inference engine can be customized to suit the nature of the application (Holsapple and Whinston, 1987). Therefore, consultation environments may range from rudimentary, rigid levels of justification to powerful, flexible ones.

User Interface

The ability to communicate with the expert system and establish a dialog that supports ease of interaction and problem-oriented flow is an essential feature of a usable system. The user interface maintains this language-focused connection to facilitate the processing of commands and the exchange of information between the user, the knowledge base, and the inference engine. The quality of the interface depends on what the user sees, what the user must know to understand it, and what actions the user must take to obtain the needed results (Turban, 1992). Several interface modes exist that determine how information is displayed, how it is entered by the user, and how well a conversational format can be maintained during consultation. A selection of the more common interface styles is presented in Table 5.4, along with a brief description of each.

Table 5.4 Common Interface Modes

Menu Interaction	Requires the user to select from a list of possible choices using keyboard or other input devices.
Command Language	Calls for the user to supply specific commands to invoke certain actions/options.
Question & Answer	Describes a machine prompting that requires the user to reply by entering keywords, menu options, or sentences.
Form Interaction	Asks the user to enter data or commands into designated spaces in a form.
Natural Language	Defines a human-to-computer dialog similar in style to a human-to-human dialog.
Object Manipulation	On-screen objects, represented as icons or symbols, are directly manipulated by the user.

Developing Knowledge-Based Expert Systems

The construction of a knowledge-based expert system follows stages similar to those in the evolution of a conventional computer program (Lein, 1989). At its most fundamental level, expert system development can be conceptualized as a six-step procedure, as illustrated in Figure 5.3. Beginning with problem identification and definition, the problem and its characteristics are determined and the requirements, resources, and goals of the expert system are defined. Once the system is specified and its purpose clearly established, the salient concepts that represent domain knowledge can be assembled into a series of rules designed to follow a clear problem-solving logic and facilitate a specific inferencing strategy. This preliminary series of rules can then be structured into a more formal knowledge base following the IF-THEN conventions addressed earlier in this chapter. At this stage in the development process, problem-specific knowledge is acquired and coded into the system. Overall, five activities are involved in this critical stage of system development:

1. **Knowledge Acquisition**—getting the knowledge from expert sources,
2. **Knowledge Representation**—placing expert knowledge into a logical structure,
3. **Knowledge Validation**—checking the correctness of the knowledge and its structure,

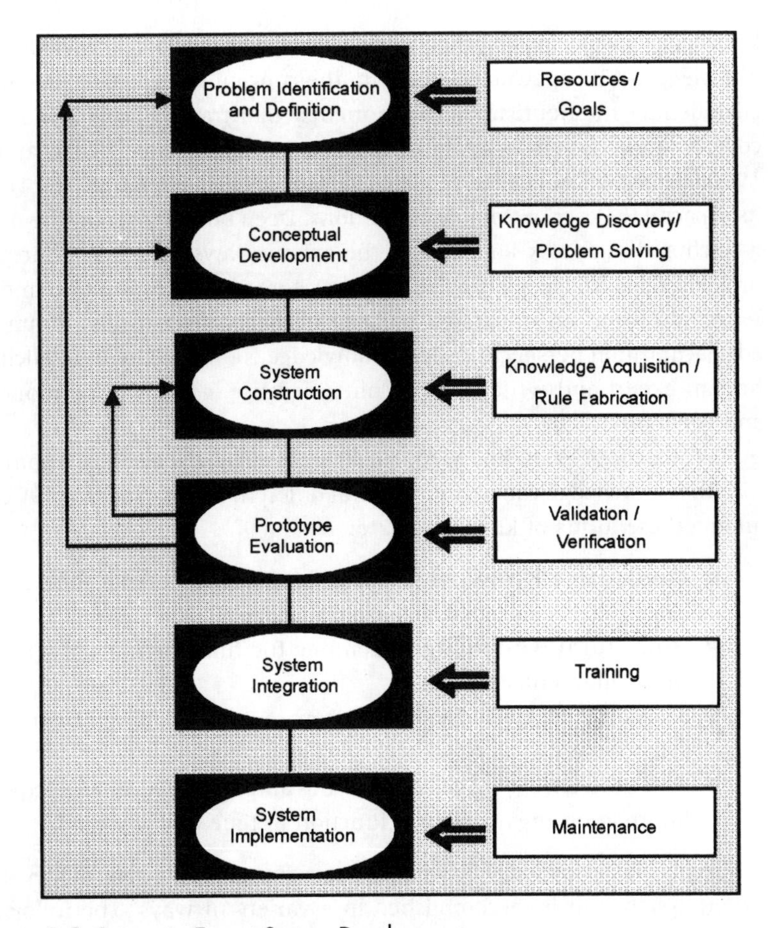

Figure 5.3 Stages in Expert System Development

4. **Knowledge Coding**—translating knowledge into a form that can be stored and processed by machine, and
5. **Explanation and Justification**—supporting the logic used to represent knowledge with clarifying information and rationalizations.

Of these five activities, perhaps none is more important and more problematic in designing a useful expert system than the knowledge acquisition phase. Knowledge acquisition is a process that involves the collection of knowledge from human experts, books, articles, documents, sensors, and other sources containing facts and relations germane to the problem. This phase directs attention to the procedures used to gather knowledge from the expert sources and the methods used to represent that knowledge (Greenwell, 1988; McGraw and Harbison-Briggs, 1989). As Turban (1992) demonstrates, the multiplicity of sources and types of knowledge required contribute to the complexity of knowledge acquisition procedures. For example, in any subject matter, knowledge can be represented at two extremes: shallow knowledge and deep knowledge. With specific reference to the concept of expertise, shallow knowledge defines the heuristics and "shortcuts" that trained professionals have discovered and learned to apply in order to perform better (Klein and Methlie, 1990). Shallow knowledge explains surface-level information and is limited in its capacity to describe complex situations. Deep knowledge, on the other hand, is anchored in theory and defines the axioms, laws, and first principles in the problem domain. Deep knowledge is therefore rooted in intuition and common sense and forms an integrated body of human consciousness. From a knowledge acquisition perspective, deep knowledge is more difficult to elicit from the human expert and write to a machine, because it tends to be more inclusive and generalized.

When considering knowledge in the abstract, it is also important to recognize that knowledge can assume several different forms (Turban, 1992). Central among these forms of knowledge are:

- **Declarative Knowledge**—describing factual relationships and descriptive representations of knowledge,
- **Procedural Knowledge**—defining the fundamentals of a process under different situations,
- **Semantic Knowledge**—characterizing the meaning of words and symbols, and
- **Episodic Knowledge**—expressing autobiographical and experimental information organized as a function of time and place.

Acquiring knowledge in these forms for the purposes of developing an expert system can be accomplished in a variety of ways. The following three common approaches have been identified (Parsay, 1985):

1. interviewing experts using a structured or unstructured interview process,
2. learning by explanation, and
3. learning by example.

In practical application, however, no single method is superior to another. Thus, in most instances, a hybrid approach is adopted (McGraw and Harbison-Briggs, 1989). Regardless of which method is employed, knowledge acquisition must achieve four basic goals (Greenwell, 1988). It must:

1. ascertain the domain characteristics of the expert,
2. investigate the analytical methods used by the expert,
3. investigate the judgmental behavior that the expert displays, and
4. categorize expertise and decide how to apply it.

In developing a knowledge-based expert system to provide advice on carrying capacity assessment, Lein (1993a) describes one hybrid approach to the problem of knowledge acquisition. In this example, an interviewing methodology was initially selected as the primary means of acquiring knowledge from the expert. However, prior to conducting a series of unstructured interviews, an extensive literature review on the topic of carrying capacity was undertaken. This search of the literature was conducted in order to uncover secondary sources of knowledge related to the problem, and it served a dual purpose. First, the review provided background information, case studies, and a listing of relevant facts and issues that had influenced research on the topic. Secondly, the background knowledge gained through the review process enabled development of details and specific questions that could be posed to the human expert during the interviewing sessions.

A series of three interviews were conducted with the human expert. The first interview involved nothing more than introducing the expert to the project and establishing a dialog for future discussions. During the second interview the expert was given a series of written questions that required answers in written form. The final interview involved a detailed examination of the written responses, permitted the expert to clarify and refine answers, and allowed the system developer to ask follow-up questions. The information, facts, rules-of-thumb, and other descriptions of knowledge acquired through the three interviews produced a body of knowledge critical to the topic that could be formalized and translated into a machine-processable form. Formalizing knowledge and placing it into a form suitable for computer manipulation are the principal tasks undertaken during the knowledge representation phase.

Knowledge Representation

A knowledge-based expert system represents the facts and relations of the problem domain as a series of production rules that take the already familiar form:

$$\text{IF}\left[\text{premise}\right]$$
$$\text{THEN}\left[\text{conclusion}\right].$$

By assembling production rules in this way, domain knowledge can be expressed in a form that simulates the reasoning process of a human expert. This approach is particularly valuable when the problem domain is a product of empirical associations developed over time through experience—a situation that aptly characterizes environmental decision making.

Coding the knowledge gained from the human expert into the production rule format involves translating the natural language concepts captured from the expert and expressing them in an unambiguous computer language that has its own well-defined syntax and semantics (Jackson, 1990). These representation languages are essentially programming languages or programming tools oriented toward organizing descriptions of objects and ideas rather than stating sequences of instructions or storing simple data elements. Examples of such languages include LISP and Prolog, while programming tools identify special-purpose expert system shells (Table 5.5). Regardless of which representation language is selected for crafting the knowledge base, a number of important issues and valuative criteria have been identified that guide the construction process (Buchanan and Duda, 1983; Jackson, 1990).

According to Buchanan and Duda (1983), a knowledge-based representation scheme must satisfy three fundamental requirements: (1) extendability, (2) simplicity, and (3) explicitness. With respect to extendability, the data structures and access programs of the expert system must be flexible enough to accommodate extensions of the knowledge base without forcing substantial revision of its initial design. Because a knowledge base tends to be built incrementally, there is a need to treat the knowledge base of the expert system as an open-ended set of facts and relations, and to keep the items of knowledge as modular as possible. Maintaining flexibility in the system also requires conceptual simplicity and uniformity. These two qualities insure that access routines can be written and modified as needed. Complex knowledge representation schemes can be difficult to modify and incomprehensible. Keeping the form of knowledge as homogeneous as possible and enforcing consistent terminology help maintain an appropriate level of simplicity. Lastly, an expert system should represent sufficient expert knowledge to give the system a knowledge base capable of high-performance problem solving. However, since the majority of knowl-

Table 5.5 Expert System Development Environments

SOFTWARE CATEGORY	DESCRIPTION
Custom-Made Systems	Application products that provide specific advice in a well-defined problem area
Shell Systems	Integrated packages in which the major components of an expert system have been preprogrammed, requiring the user to insert a knowledge base
Support Tools	Specialized software packages to assist in developing one or more components of the expert system
Hybrid Systems	Systems composed of several support tools and programming languages that permit rapid prototyping
Programming Languages	Specific computer programming languages designed to support AI applications or a standard procedural language

edge bases are built incrementally, providing a means for inspecting and debugging them easily is a necessity. With items of knowledge represented explicitly and in comparatively simple terms, the developers of the knowledge base can ascertain which items are present and (by inference) which are absent.

In addition to these three criteria, several other factors may influence the design of a knowledge base, including:

Consistency—A desirable quality, though difficult to achieve. Since much of the knowledge entered into the knowledge base is previously uncodified and somewhat uncertain, rules may conflict or suggest incompatibility.

Syntactic Completeness of the Representation Language—A logical requirement in knowledge-based systems that many fail to satisfy. Syntactic completeness involves a check of assertions and other qualified statements in the knowledge base to insure that their expression leads to a logical conclusion or valid interpretation of the evidence.

Semantic Completeness of the Knowledge Base—Because of the nature of the knowledge base and the way it is constructed, it is possibile that some aspect of the problem will be omitted. Here, concern focuses on the completeness of the rules and the inferences made by the system.

Precision—In some specialized domains, precision may be determined for many of the facts and rules, but not for all. Since there is a temptation to make overly precise assertions even though there is no justification for them, representing degrees of imprecision is an important feature of a representation methodology.

Default Knowledge—An essential feature that protects against incompleteness. In a rule-based system, each class of actions can have an explicitly stated default that can complete an inference chain in the absence of contrary evidence.

Assessing a knowledge representation schema introduces three additional valuative criteria worth noting: logical adequacy, heuristic power, and notational convenience (Jackson, 1990). Briefly, logical adequacy implies that the representation is capable of making all the distinctions needed to address the problem. Heuristic power suggests that given the structure of the representation language, there must be some means of using that representation to solve problems. Notational convenience relates to the complexity of the representational language and supports the idea that the language or tool chosen is not overly complicated. This is an important consideration given the fact that substantial amounts of knowledge may require encoding in order to create an adequate knowledge base. Therefore, the expressions produced during knowledge representation should be relatively easy to read, write, and understand without our knowing how they will be interpreted by the computer (Jackson, 1990).

Illustrating the Design Model

The components of an expert system discussed above are elements that come to life only when connected to a structure that provides an underlying logic that emulates a human reasoning process. Thus, as with any information technology, it is possible to have all the pieces and still not support decision making. Pieces need glue in order to stick together. With respect to a knowledge-based expert system, that glue comes from the proper application of good reasoning techniques to a large store of problem-specific knowledge. Designing an expert system is, in essence, capturing a reasoning technique or style and applying it to a knowledge base that has been crafted into a framework that reflects the logical associations implied by the selected reasoning style. A reasoning method cannot be completely independent of the class of problems they must solve, yet must not be so specialized that they become inapplicable elsewhere (Weis and Kulikowski, 1984). Within the context of environmental analysis and decision making, a type of problem frequently encountered involves: (1) identifying conditions or qualities from a set of measures or criteria, or (2) selecting among the qualities or conditions based on those measurement criteria. An example of this type of problem is deciding what a parcel of land can be used for given its physical and locational characteristics, or determining if a measured value exceeds a set standard or indicates a prespecified level of compliance. Examples of these types of questions fall within the general class of consultation or classification tasks characteristic of interpretive and diagnostic problems and encourage the development of diagnostic expert systems to assist in environmental decision making (Lein, 1989; Lein, 1990).

Designing a diagnostic expert system based on the production rule paradigm, while an effective way of expressing inferential knowledge well suited to environmental applications, still demands a logic framework that facilitates inference using a complex system of rules. According to Weiss and Kulikowski (1984), one framework that provides an underlying logic structure to guide the application of production rules is the classification model. A diagnostic or consultative expert system designed to implement the logic of the classification model has the primary task of selecting a conclusion from a prespecified list of possibilities. In the abstract, this quality suggests that there are three interrelated lists in the knowledge base:

1. a list of possible observations,
2. a list of possible conclusions, and
3. a list of rules relating observations to conclusions.

When assembled into the knowledge base, the expert system will resemble the design illustrated in Figure 5.4, with its knowledge base organized to represent the classification logic specific to the problem domain.

The design of an expert system using the classification model for diagnos-

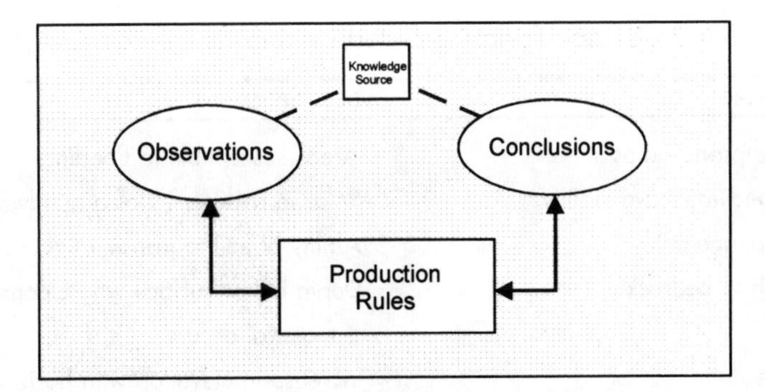

Figure 5.4 The Classification Model (Based on Weiss and Kulikowski, 1984)

tic or interpretive consultation is best explained by means of an example. The example selected is an extension of the work first introduced by Lein (1990). In this demonstration of consultative expert systems, a knowledge-based expert system was described that provides advice on the developmental suitability of land for a selection of land uses. The knowledge base developed for this domain drew heavily on the general knowledge and rules-of-thumb guiding an expert's interpretation of developmental suitability and constraint. To produce this knowledge base and obtain an expression of suitability sufficient to guide the decision maker on the possible uses of a land parcel, information on the following considerations had to be acquired and coded into the knowledge base:

1. a set of criteria that effectively communicated constraint,
2. a strategy that enabled the system to evaluate constraint for the chosen land use,
3. a defined set of limitations that categorized each criterion, and
4. an estimate of confidence that supported the conclusion arrived at by the system.

The landscape factors selected for inclusion in the knowledge base are provided in Table 5.6. They were selected based on a threefold rationale. First, they defined landscape qualities known to be crucial in evaluating developmental suitability. Secondly, they facilitated categorical expressions that could communicate constraint over a range of factor levels. Lastly, they represented landscape information commonly recorded and stored in database files housed in geographic information systems.

Because each of the factors listed in Table 5.6 can exert contrasting levels of importance relative to the expression of suitability, a semantic partition that separated land uses according to their scale was employed in the knowledge base (Lein, 1990). This device permitted a subjective determination of the project's scale to direct the inferencing process and provided a more logical connection between the type of development and the categorical expression used

Table 5.6 Landscape Factors Selected for Knowledge-Base Development

FACTOR	DESCRIPTION
Flood-prone areas	Areas susceptible to flooding
Groundwater availability	Relative availability of groundwater
Soil permeability	Ability of soil to transmit water
Depth to bedrock	Depth below surface where consolidated rock material occurs
Slope	Increase in elevation with horizontal distance
Drainage class	Rate at which water is removed from soil
Shrink-swell potential	Relative change in volume with changes in moisture content
Corrosion potential	Soil-induced chemical action
Erosion potential	Measure of potential soil loss
Frost action	Freezing effects on soils
Soil texture	Measure based on soil particle size
Available water capacity	Ability of soil to hold water
Depth to seasonal water table	Indicator of potential soil saturation
Bedrock geology	Bedrock type at depth below surface

to evaluate constraint. In the present example, land uses were divided into either small-scale or large-scale development classes. A series of rules were constructed based on land use type and project scale. These two distinctions enabled a matching of scale with the appropriate valuative criteria and greatly simplified inferencing.

With the general architecture of the knowledge base described according to the scale-criteria logic explained above, knowledge representation concentrated on designing rules that characterized each factor as an expression that explained how it influenced the suitability of the site at a given scale of development. This required placing factor levels in categories closely corresponding to the classification scheme used in map-based suitability assessments. By conforming to established conventions, rules could be constructed to relate a quantity at a given level to a use at a specific scale and ultimately to a statement of limitation (conclusion). The expression of constraint stated in a rule's conclusion was also treated categorically. This allowed the system to communicate varying degrees of restrictions using natural language concepts enhancing the evaluation of trade-offs among offending site characteristics.

The final stage in developing the knowledge base in this example involved translating the interpretive knowledge of the suitability assessment problem into the IF-THEN format of the production rule. The rules created during this phase defined a premise that was divided into one or more conditions and a conclusion. In addition, a certainty factor was assigned to each rule to express the degree to which the facts expressed in the premise supported the conclusion as given by the system. To complement each rule and provide justification and guidance to the user of the system during a consultation, explanatory notes were included with each rule (Figure 5.5).

A sample of the rules developed for the suitability assessment knowledge base showing the general structure of the design model according to the classification paradigm follows. The rules presented are limited in scope for a single land use type and depict the logic followed to establish the degree of soil limitation. They are not presented here as a fully developed knowledge base, but merely to illustrate production rule format and the basic flow of a simple line of reasoning. Therefore, to determine suitability, land use type (single-family dwelling) is compared to an environmental factor, and a conclusion is obtained relating the potential impact the landscape factor may impose. The physical knowledge base evaluating this condition may take one of the following forms:

Rule Number: 1
IF:
Proposed use is single-family dwelling
AND Shrink-swell potential is moderate
THEN:
Soil limitations are moderate.

RULE NUMBER: 56

IF:

Proposed use is industrial
or large commercial structure
AND bedrock geology is dolomite
AND bedrock condition is solution-weakened

THEN:

bedrock stability limitations are moderate
Probability = 8/10

NOTE: Not all types of bedrock can provide a sound foundation for large structures. Certain rock types such as limestone or dolomite are soluble in weak acids found in groundwater. Therefore, such rock types are subject to possible weakness.

Figure 5.5 General Format of a Production Rule

Rule Number: 2
IF
Proposed use is single-family dwelling
AND septic system is yes
AND permeability is 2.0 to 0.20 inches/hour
THEN:
Soil limitations are moderate.
Rule Number: 3
IF:
Proposed use is single-family dwelling
AND shrink-swell potential is very high
THEN:
Soil limitations are severe.
Rule Number: 4
IF:
Proposed use is single-family dwelling
AND septic system is yes
AND permeability is 20.0 inches/hour or greater
THEN:
Soil limitations are severe.
Rule Number: 5
IF:
Proposed use is single-family dwelling
AND erosion potential is greater than 4 tons/acre/year
THEN:
erosion limitations are severe.

To gather information in order to invoke the rules, the system queries the user through its user interface facility. Query can take several forms, including:

1. entering options from a menu,
2. pointing to an icon,
3. pointing to a menu selection, or
4. entering specific answers to questions that appear in dialog windows.

In the present example, questions or options were posed to the user, who was required to respond during the consultation via the keyboard. This mode of interface provided needed data input to the system, which prompted the inferencing process. The questions posed were linked to the main qualities or conditions contained in a rule's premise. The questions represented in the knowledge base conformed to a standard style. Here is possible syntax demonstrating this form:

Question shrink-swell: "Would you characterize the soil shrink-swell potential as moderate, slight, or severe?"

Question erosion: "Please give me an estimate of possible soil loss at the site (in tons/acre/year)."

Question septic: "Is a septic system planned for the site (yes or no)?"

The specific format adopted to guide the design of an expert system is generally a function of the development environment or the representational language that has been selected by the system designers. As noted previously, an expert system can be physically coded using an AI language such as LISP or Prolog, or by means of an expert system development tool (often referred to as an expert system shell package). The merits of each method are discussed in the following section.

The Design Environment

An expert system has been defined as a software environment consisting of a knowledge base, an inference engine, a user interface, an explanation facility, and a knowledge acquisition system. Assembling these elements into a workable design requires time, money, personnel, and hardware. It is also a process that is strongly influenced by the design environment. Expert systems can be constructed either by using a programming language or by means of an expert system shell. Programming languages, while offering flexibility, typically require the developer of the system to design the knowledge base and implement an inference engine that can access knowledge. As a consequence, development often takes longer, although the resulting expert system may more closely match the needs of the problem domain. However, the relative attractiveness of AI programming languages is diminishing as more features are added to the shell environment.

An expert system shell is a complete expert system minus the knowledge base. A typical shell package contains an inference engine, a knowledge acquisition facility, an explanation facility, a user interface, and a knowledge-base management system. With an expert system shell, the system developer can concentrate efforts exclusively on creating the knowledge base. Thus, by using the shell approach, expert systems can be built faster and require less specific programming skills on the part of the developer. These two factors combined contribute to a significant reduction in the cost of the design process and make shell packages an attractive method of expert system design.

In general, an expert system shell performs three major functions (Parsaye and Chignell, 1988):

1. it assists in building the knowledge base by allowing the developer to insert knowledge into a predetermined format,
2. it provides methods of inference or deduction that reason on the basis of the information contained in the knowledge base, and
3. it offers an interface that permits the user to set up a reasoning task and to query the system about its reasoning strategy.

For these three reasons, shells are employed extensively both for the scoping-out of an expert system application and as a starting tool in the development of prototype systems. Shells are not a panacea, however. Several limitations of shell packages can be noted, including: (1) inflexibility, (2) poor documentation, and (3) lack of completeness. Consequently, selecting the correct problem scope and choosing the right development tool for constructing the expert system are two of the more difficult decisions to make when considering the expert system approach (Waterman, 1986). Waterman (1986) offers the following six guidelines to help in selecting the appropriate expert system design environment:

- Does the tool provide the developer with the required power and sophistication?
- Are the support facilities adequate considering the time frame for development?
- Is the tool reliable?
- Is the tool maintained by its developers?
- Does the tool have the features suggested by the needs of the problem?
- Does the tool have the features suggested by the needs of the application?

These guidelines are important in the successful development of an expert system, but another (often overlooked) aspect of the development problem is the concept of uncertainty. Specifically, attention should be given to the capacity of the development tool to provide mechanisms for incorporating uncertainty and for expressing estimates of confidence regarding a conclusion reached by the system during consultation. Because all reasoning and decision making take place in an atmosphere of uncertainty, an expert system designed to mimic a human expert must be able to reason with uncertain information and communicate certainty or confidence in its conclusions just as its human counterparts do. Several of the more common methods for incorporating uncertainty in a knowledge base and using it as a basis for inference are discussed in the following section.

Incorporating Uncertainty

Coping with uncertainty, or reasoning with uncertain or inexact information, is a complex area in the field of artificial intelligence. Detailed treatments of the topic can be found in DeMantaras (1990), Sombe (1990), Klir and Folger (1988), and Castillo and Alvarez (1990). Because of the approximate nature of environmental processes and the activities that describe environmental decision making, a knowledge base and reasoning procedure cannot be created under the assumption of absolute certainty and still be of assistance to decision makers. Therefore, to model environmental situations realistically, the knowledge representation and inferencing methods selected to guide creation of an expert system should be supplemented with procedures for handling uncertainty.

Presently, there are two general ways in which uncertainty may develop an expert system (Parsaye and Chignell, 1988). In knowledge-based systems, one source of uncertainty is the incompleteness or unavailability of information. A very different source of uncertainty is the inherent imprecision of the knowledge itself. Because expert systems assume that their knowledge base is complete, the treatment of uncertainty typically focuses only on the problem of imprecise information. Several approaches have been developed to manage uncertainty in this context, and each have their relative advantages and disadvantages. Four of the more widely applied methods for representing uncertain information in an expert system are based on:

1. Bayesian Probabilities,
2. Certainty Factors,
3. Dempster-Shafer Theory of Evidence, and
4. Fuzzy Logic.

Each of these approaches is reviewed briefly below.

Bayesian Methods

The Bayesian approach to uncertainty representation is based on the supposition that the prior probability of an event should be incorporated into the interpretation of a situation. According to this method, the Bayes Theorem provides a mathematical model for reasoning by combining prior beliefs or an event or fact with evidence to form an estimate of uncertainty. The theory is predicated on three defining assumptions: (1) mutually exclusive hypotheses, (2) conditionally independent evidence, and (3) completely enumerated sets of hypotheses.

Symbolically, the Bayes Theorem can be defined as:

$$p(H/E) = p(H) * p(E/H)/p(E) \tag{5.1}$$

where E represents the evidence and H represents a set of hypotheses. In the practical application of this approach, existent evidence (E) is typically expressed as a subjective probability. Although problematic and controversial, the interpretation of subjective estimates of uncertainty as if they were mathematically defined probabilities rests on the assumption that under most conditions, they express fairly accurate predictions of events. This point is well argued by Lindley (1982) and effectively demonstrated by Mann and Hunter (1988).

As a method for managing uncertainty, the Bayesian approach is tailored to well-structured situations in which all the data is available and the assumptions that guide its application are satisfied. However, this procedure becomes more difficult to implement as the knowledge base grows in size since it becomes functionally impossible to change one probability expressed in a rule without causing an intractable ripple effect. Expert systems that employ Bayesian reasoning inevitably violate the mathematical principles used to derive it, which led Parsaye and Chignell (1988) to conclude that Bayesian probabilities are better suited as methods for combining inexact values than as a probabilistic approach to uncertainty.

Certainty Factors

While Bayes Theorem captures the notion of uncertainty effectively, its accurate use depends on knowing many probabilities. An alternative way of measuring uncertainty in a rule is based on Certainty Theory. Certainty Theory employs the concept of a certainty factor to express the degree of belief in an event based on evidence. One means of expressing certainty in a knowledge-based expert system is to use a simple 0–100 scale. In this simple scaling convention, the value 100 represents absolute truth or complete confidence in a conclusion and the value zero represents absolute falsehood. Although these values are not probabilities, they are designed to communicate a subjective level of belief in the premise of a rule, or the conclusion it explains. For instance, in a rule evaluating slope constraints, a certainty factor (*CF*) may be expressed as follows:

IF slope exceeds 15 percent,

THEN slope constraint is moderate, $CF = 80\%$.

In this example, the rule can be interpreted simply as follows: If a land area is characterized by a 15 percent slope, then there is an 80 percent (subjective) chance that slope will constrain potential development. With this nonprobabilistic approach, the certainty factor (moderate = 80) identifies a belief based on evidence, experience, or opinion that slopes of 15 percent introduce moderate constraint. As is the case in this approach, the certainty factor expressed in the rule does not have to total 100.

A more formal method used to derive certainty factors introduces two additional measures. These measures, symbolized as *MB* and *MD*, respectively, measure the relative decrement of disbelief in a hypothesis given some evidence, and the relative decrement of belief in a given hypothesis due to some evidence. Both measures are scales along the interval [0,1]. Thus, in the previous example that evaluated slope constraint, if an expert believes that there is an 80 percent (0.80) chance that a 15 percent slope introduces a moderate constraint, then there is also a 20 percent (0.20) chance that it does not. Following this logic, the certainty factor (*CF*), combines the measures of belief (*MB*) and disbelief (*MD*) according to the formula

$$CF[P,\ E] = MB[P,\ E] - MD[P,\ E], \tag{5.2}$$

where

$$P = \text{probability}$$

$$E = \text{evidence.}$$

Using this general relationship, certainty factors can be employed to combine different estimates of experts in one rule or in two or more rules. For the single-rule example, where each clause in the premise must be true, this relationship becomes

$$CP(A,\ B,\ \text{and}\ C) = \text{minimum}\left[CF(A),\ CF(B),\ CF(C)\right]. \quad (5.3)$$

Where only one clause in the premise must be true, the certainty factor is determined by

$$CF(A\ \text{or}\ B) = \text{maximum}\left[CF(A),\ CF(B)\right]. \quad (5.4)$$

If a knowledge base consists of several interrelated rules, and each rule has the same conclusion but expresses a different certainty estimate, the general logic for combining confidence is as follows:

$$(\text{Rule 1} + \text{Rule 2}) - (\text{Rule 1} * \text{Rule 2}). \quad (5.5)$$

Recently, Lein (1993a) employed certainty factor algebra for conjunctions (AND) based on the assumption that a chain of statements connected by AND will be only as strong as its weakest link. In the demonstration system, this relationship translated to:

$$CF(A\ \text{and}\ B\ \text{and}\ C\ \text{and}\ D) = \min(CF(A),\ CF(B),\ CF(C),\ CF(D)). \quad (5.6)$$

Theory of Evidence

Similar in concept to Certainty Theory, the Dempster-Shafter Theory also applies the concept of a belief function to distinguish between uncertainty and ignorance in the knowledge base. According to the Dempster-Shafter Theory, a belief function models uncertainty by means of a belief measure that expresses the degree of belief in a conclusion given the evidence. This measure, when applied to a rule in the knowledge base, communicates a level of confidence in the range 0 to 1, where 0 indicates no confidence and 1 explains absolute confidence.

When compared to traditional probabilistic approaches, the Dempster-Shafter Theory captures the natural behavior of reasoning by narrowing the hypothesis set down to a smaller number of possibilities as the evidence increases. Therefore, belief measures can be attached to a chain of events or facts in the knowledge base and then combined to form a single line of reasoning. The belief measures can then be summed to yield an overall expression of certainty according to the relation

$$\text{Bel}(X) = \sum m(Y), \tag{5.7}$$

where m represents the measure of belief for (Y) rules.

In a knowledge-based expert system, this "global" belief can be expressed as

$$\text{Bel}(\text{conclusion, premise}) = m(\text{clause } 1) + m(\text{clause } 2) + \ldots + m(\text{clause } n),$$

where m expresses the measure of belief for each condition expressed in a given rule's clause.

Fuzzy Logic

The importance of fuzzy logic in decision making was discussed at length in Chapter 2. Applying those concepts as a means of representing uncertainty relies chiefly on developing techniques to represent fuzzy knowledge and adapting these techniques to the structure of a knowledge-based expert system. As shown by Turban (1992), a production rule has no concrete effect unless the data completely satisfies the antecedent of the rule. Should two rules fire simultaneously, the system's conflict resolution strategy is invoked to try to manage this ambiguity. A rule-based system employing fuzzy logic will execute all its rules during each pass through the knowledge base. However, because the rules are fuzzy, they produce contrasting effects on the chain of reasoning. The strength of these effects depends on the relative degree to which their fuzzy antecedent propositions are satisfied by the data. If the match between data and rule is "perfect," the result of the rule firing is an assertion that agrees with the consequent proposition of the rule. If the antecedent is only partially satisfied, the match between data and conclusion is made in proportion to the fuzziness of the match as defined by its membership function. A detailed treatment of fuzzy rules, the assignment of membership grades, and their combination as a means of representing uncertainty can be found in Giarratano and Riley (1989).

Environmental Applications

Within the last decade there has been a steady interest in developing expert systems for specific applications in environmental management and planning. The evolution of environmental expert systems is well described by Davis and Clark (1989) and more recently by Plant (1993) and Warwick et al. (1993). Overall, a cursory review of the literature will reveal a range of expert systems developed to assist in environmental impact assessment (Lein, 1989; Geraghty, 1992; Mercer, 1995), environmental evaluation (Julien et al., 1992), land-degradation and carrying-capacity assessment (Balachandran and Fisher, 1990; Lein, 1993), suitability and site selection (Lein, 1990; Edamura and Kawai, 1991), planning (Han and Kim, 1989; Yan et al., 1991; Davis et al., 1987), and decision support (Davis and Grant, 1987; Han et al., 1991; Armstrong et al., 1990).

In the example listed above, the expert systems can range from demonstration and proof-of-concept systems to working prototypes. Regardless of their apparent status, the systems selected in this sample illustrate how expert systems can be developed to assist in environmental decision making, and demonstrate the breadth of problems expert systems can address. For the most part, today's environmental expert systems fall into four general application categories: (1) interpretive systems, (2) planning and assignment systems, (3) predictive or prescriptive systems, and (4) diagnostic systems. Although other types of system applications are possible—notably, applications involving monitoring and instruction—the candidate situations well documented in the literature are those that

- occur often,
- are complex,
- require expert knowledge,
- involve some degree of uncertainty,
- are dynamic, and
- demand consistent responses.

One emerging application area that holds promise involves linking the knowledge-base environment of an expert system to the database environment of an information system (that is, GIS). Coupling the two technologies creates an "intelligent" information system and places the expert system in the role of an intelligent front end to database technologies and mapping systems. The possibilities of such hybrid systems have been actively explored for several years. Indeed, interest in linking expert systems and GIS can be traced to the works of Gardels (1987), Morse (1987), Kofron (1989), and King (1990). This early research culminated in the development of several demonstration systems in application areas such as municipal site planning (Maidment and Djokic, 1990), solid waste landfill site selection (Davies and Lein, 1991), forest soils mapping (Skidmore et al., 1991; Skidmore et al., 1996) biophysical landscape classification (Coughlan and Running, 1989), and zoning (Chen et al., 1994).

The physical coupling of an expert system with a GIS has been approached in several ways. Presently, four primary interfacing strategies can be identified (Fischer, 1994; Lein, 1992):

- **Full Integration**—an interface strategy where expert system tools are embedded into the GIS as specific operations,
- **Loose Coupling**—an approach that enables communication between the expert system and the GIS, which may also include simultaneous operations,

- **System Enhancement**—a method of integration that gives the expert system database functionality or gives the GIS rule-base capabilities, and
- **Tight Coupling**—a technique that employs one technology to control operation and execution of the other.

Implementing any of these interfacing strategies depends heavily on the approach or philosophy underlying the development effort. Lein (1992b) notes four main expert-GIS development paradigms that more clearly define the relative role these systems play in a combined application. These include:

1. the intelligent user interface,
2. expert control of GIS functionality,
3. GIS operation with expert system assistance, and
4. expert system processing an interpretation with GIS display.

The relative merits of each are discussed below.

Placing the expert system in the role of an intelligent front end creates an interface style where the expert system functions largely as an assistant that performs critical database searches and guides the user through specialized data manipulations. The expert system contains domain knowledge that instructs the system as to which data items to access, which calculations to make, which items to process, and how to perform and display problem-specific searches. But it does not perform analytic operations using the GIS or toolbox. Instead, the knowledge base conducts its analysis directly from the database and returns the appropriate answers to the decision maker.

Designing an interface based on the paradigm where the expert system controls GIS functionality or where the GIS calls on expert guidance requires a high level of openness between the two software systems. Openness in this context means that the systems permit full integration of all external routines within their main kernels. An open design is essential for this purpose since it facilitates development of interface routines that permit simultaneous operations and intersystem communication. Presently, the necessary level of openness between proprietary software systems does not exist. As a result, full integration of commercial expert system packages with their commercial GIS counterparts is a vexing proposition. The final design paradigm identified by Lein (1992b) places the expert system in the role of a knowledge-driven processor. Here, the expert system performs classification and selection functions according to its problem domain: reading data, making comparisons, and selecting appropriate combinations of data items to satisfy its goal. The results obtained via these procedures are then passed on to the GIS for display or further analysis.

Although the implementation and configuration of a knowledge-based geographic information system can vary, the intent guiding their creation remains consistent: to utilize expert domain-specific knowledge to improve the transformation of data into more meaningful information products. Realizing this goal also enhances the utility of spatial databases, particularly in situations where data may be available to solve problems, but expertise is not. An excellent example of the potential of an expert-GIS system designed to assist in natural resource management and decision making is the IRMA system described by Loh and Rykiel (1992).

The Integrated Resource Management Automation System (IRMA) is a computer-based set of interacting software applications that provides intelligent decision support in a specific problem context. In its present implementation, its knowledge is limited to forest planning. Within this domain, applications exchange data through a user interface that regulates interaction and insulates the decision maker from the technical details of the system. The IRMA system, as described by Loh and Rykiel (1992), consists of five subsystems (listed below), each with its own specific functionality:

1. a user interface shell that acts as the "switchboard" for the systems, coordinating how the remaining IRMA components interact,
2. a database management system that provides access to and manipulation of a relations database through an SQL message interface,
3. a geographic information system that provides capabilities for spatial data analysis,
4. a rule-based reasoning system that consists of a knowledge base and an inference engine, and
5. a data exchange system that performs data import and export functions.

As an example of an integrated information technology designed to support environmental decision making, the IRMA system creates a machine environment where management opportunities can be identified, options can be evaluated, and the best alternative selected automatically. Although the IRMA system described by Loh and Rykiel (1992) is a prototype that lacks the level of detail necessary for practical implementation, the system illustrates the value of integrating technologies such as expert systems, GIS, and simulation models to create a new generation of intelligent automated problem solvers. This potential, however, is tempered somewhat by several unresolved issues that surround expert systems design.

Design Limitations

The growing number of applications of expert systems beyond the research laboratory has placed new emphasis on three essential yet frequently overlooked aspects of this technology: (1) validation, (2) verification, and (3) adaptability. As with any information technology, there is a tendency to view an expert system as a black box and to assume that because the system resides within a computer, it is necessarily correct and reliable. With knowledge-based systems, such an assumption fails to take into consideration the fact that the expert system performs at two different levels that greatly influence its operation. One level is the machine environment. This level defines its installation and how the system runs off of the computer. The second level is its expert performance—that is, the quality of the knowledge the system contains and the advice the system provides during a consultation. Because an expert system can execute flawlessly at the machine level yet provide useless answers at the expert level, rigorous testing and critical evaluation are required before any system can be implemented.

Validation

Evaluating the performance of an expert system relies on careful validation and verification. Validation of an expert system is the establishment of the quality of system performance relative to that of a human expert in the same domain. Through validation the attempt is made to substantiate the accuracy of the system and determine that the system performs at an acceptable level of accuracy. Validation typically treats the expert system as a black box and does not assess its internal structure. Tests are performed using the system to determine whether it satisfies a set of accuracy requirements (Suen et al., 1990). In other words, the system is given a series of test problems and its responses to those problems are analyzed. Several questions may be raised during validation testing. These have been summarized by O'Keefe et al. (1988) and include concerns as to:

- what to validate,
- what to validate against,
- what to validate with,
- when to validate,
- how to control for bias, and
- how to cope with multiple results.

Answers to these questions are provided by O'Keefe et al. (1988) and are summarized in Table 5.7.

In general, the main considerations when developing a validation test include concerns regarding the selection of test cases, the evaluations of test responses, and the role of statistical analysis (Suen et al., 1990). These concerns are summarized in Table 5.8.

Verification

Verification is the process of substantiating the correct implementation of a system's specifications. It is a guarantee that an expert system will behave as required (Wood and Frankowski, 1990). As a performance test, verification treats the expert system as a "glass box" and requires access to the knowledge base and inference engine. By careful testing of the knowledge base, incorrect or inconsistent rules can be detected and flaws in logic identified. Generally, verification involves two related checks (after Suen et al., 1990):

Table 5.7 Guidelines for Validating Expert System Performance

Rule 1: Validate systems only against an acceptable performance range for a prescribed input domain.
Rule 2: Validation should be part of the development cycle where cross-sectional performance validation is conducted prior to implementation.
Rule 3: The risk of using invalid systems (user's risk) must be considered in relation to the risk of not using valid systems (builder's risk).
Rule 4: Selection of an appropriate qualitative method is critical. This may be field testing, Turing tests, subsystem validation, or sensitivity analysis.
Rule 5: Ultimate validation must be quantitative; however, quantitative methods must be used only where applicable and should be used as informally as possible.

1. **Structure Checking**—to find inconsistencies, redundancy, subsumptions, and cyclic dependencies, and
2. **Semantic Checking**—to locate range errors, cardinality errors, illegal values, and incorrectness in rules.

An excellent overview of knowledge-base verification can be found in Constantine and Ulvila (1990) and Lunardhi and Passino (1995).

Adaptability

Evaluation of expert systems designed for environmental domains should also include adaptability as a performance criterion (Dyer, 1989). Because a knowledge base is highly portable, it can be easily shared and transported to other geographic locations for application in the local environmental decision-making process. This important quality of an expert system places a premium on the adaptability of its knowledge base. Adaptability in this context defines the ease with which an expert system designed for a particular geographic location can be applied to another location without significant deterioration in its performance. With a thorough analysis of the knowledge base, when rules are discovered that display geographic dependence, it may be possible to introduce greater independence by altering the method of knowledge representation or

Table 5.8 Methods Used for Validating Expert Systems

1. Test Cases

- Should be based on actual situations
- Should cover a range of levels of difficulty
- Should be generated by unbiased experts
- Should test as many aspects of the system as possible
- Should be conducted in the field

2. Evaluating Test Responses

- Must be free of bias, prejudice, and parochialism
- Must be obtained from experts in the field

3. Establishing Performance Measures

- Based on the size of the knowledge base
- Based on the number of facts and rules
- Based on the depth of the knowledge base (contains a high proportion of rules to facts)
- Based on the breadth of the knowledge base (contains a high proportion of facts to rules)
- Based on the time required to solve a typical problem
- Based on the difficulty of problems that can be solved

the method of data entry, or by using templates that encode knowledge in a more standardized form (Dyer, 1989). Perhaps the most promising way to achieve increased knowledge-base adaptability is through a more in-depth knowledge acquisition system coupled with a knowledge-base architecture that separates deep knowledge from the superficial knowledge represented by heuristics (Dyer, 1989).

With careful validation and verification of the expert system, an important source of knowledge that can greatly assist in decision making will be gained. By paying close attention to adaptability issues, we may be able to produce a system that can be transported to another geographic location and used with confidence to provide needed expertise where it is lacking. There are, however, alternative methods of representing knowledge for application in a computer environment. One promising alternative, the artificial neural network, is explored in Chapter 6.

Neurocomputing and Neuro-Decision Making

Decision making under conditions of uncertainty, particularly when the data on which the decision is based is imprecise or poorly connected to a predictable outcome, has encouraged the development of techniques and strategies to reduce the impact of uncertainty or incorporate it explicitly into the decision-making process. With data-driven systems, such as GIS or model-based systems, minimizing uncertainty directs efforts toward the issues of data quality, error propagation, systems representation, and systems modeling. With respect to knowledge-based systems, uncertainty is expressed as a function of knowledge stated explicitly in the form of rules that are composed into a knowledge base and combined with an inferencing strategy. There are other methods as well for representing knowledge and modeling the uncertainty inherent in decision making. An alternative means of characterizing problem-specific knowledge, data, and uncertainty that has become popular again is the use of an artificial neural network. In this chapter, the concept of an artificial neural network is introduced and the role of neural computing in environmental decision making is examined.

Neural Network Principles

An artificial neural network takes its name from the biological network of neurons that comprise the human brain. Artificial neural networks are biologically inspired, but are a much more simplified abstraction of their biological counterparts. As a stylized model of the human brain, an artificial neural network is a highly interconnected network of simple computational elements that process information according to its dynamic-state response to a set of external stimuli (Nelson and Illingworth, 1991). As an information technology, an artificial neural network attempts to replicate artificially many of the human brain's neural functions associated with thinking (Kohonen, 1988a). Of the exclusively neural functions replicated by artificial neural networks, the implementation of artificial sensory functions and functions related to motor control are probably the most widely developed. To parallel these actions and simulate the functionality and decision-making processes of the human brain, an artificial neural network model represents knowledge implicitly as a function of its network topology coupled with the processing paradigm guiding its design and implementation. Therefore, a typical artificial neural network consists of a predetermined number of simple, highly interconnected nodes that process information according to an input stimulus and generate an output that is transmitted throughout the network.

The basic design of an artificial neural network, stressing its topological structure, is presented in Figure 6.1. The design elements central to this form of information processing are the nodes that form the connecting points in the network. According to neural network theory, each node in the network functions as an independent information processor and is referred to as a processing element. These elements interpret multiple-input signals and combine them into a single output just as a biological neuron produces an output based on the input stimuli it receives. Thus, to gain a better understanding of artificial neural networks (ANN), their design, and their mode of information processing, we need to refer to the biological prototype.

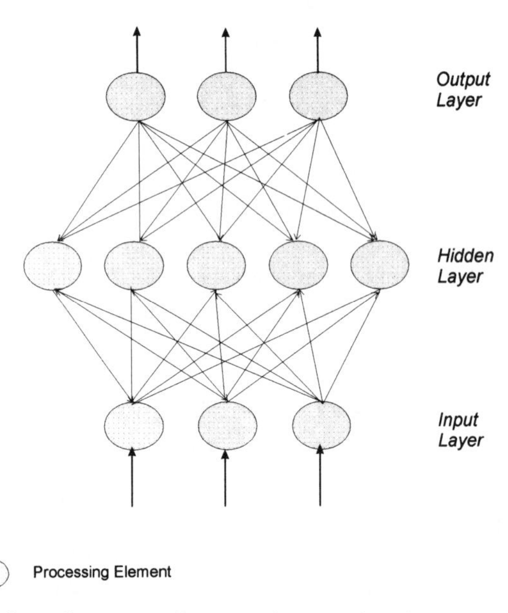

Figure 6.1 General Configuration of a Neural Network with One Hidden Layer

The Biological Connection

The artificial neuron, serving as the backbone of the ANN, was designed to mimic the first-order characteristics of the biological neuron (Wasserman, 1989). According to our present understanding of the human brain, a biological network of neurons is a system of electrically active cells that interact with one another through the flow of local ionic currents. These currents, as Cruz (1989) explains, travel along fibers that connect cell bodies into networks that form nerve structures. The input fibers leading to a cell body are called dendrites, while the output fibers are called axons. A dendrite's primary function is to conduct impulses toward the cell body, while axons function to conduct impulses away from the cell body (Nelson and Illingworth, 1991). By means of this connection scheme, each fiber transmits pulses of current in one direction, which suggests that neurons feed information to one another in definite patterns.

The gap between two neurons defines a neural pathway and describes the point where the termination of the axon from one neuron comes close to the cell body or dendrite of another. This point of near confluence is called the synaptic gap, When a nerve impulse reaches this gap, it causes the release of a neurotransmitter, which diffuses across the synaptic gap. As this chemical is received at the dendrite end of the adjoining cell body, it generates a nerve impulse that flows along the dendrite of this target cell. There, the inputs are weighted and the resulting quantities are summed. If the combined current exceeds the threshold for that neuron, the neuron fires, producing a signal as output that propagates down the network.

The generalized neural signaling scheme has been effectively summarized by Cruz (1989) and Simpson (1990) and includes the following steps:

- source nodes send out impulses along their axons,
- at the termination of an axon, the nerve impulses cross the synaptic gap and enter the input fibers of sink cells,
- the resulting nerve impulses travel along their dendrites and arrive at the target cell,
- each cell body continually integrates the currents that arrive via its dendrites,
- the summed current is compared against a "firing threshold," and an output pulse train is produced if the threshold is exceeded, and
- the outgoing impulse continues this cycle down the network.

Based on this procedural description, the following important qualities of a biological network can be noted that heavily influence the design of an artificial neural network:

1. each neuron is a simple, independent processor, receiving, combining, and producing impulses as an autonomous information-processing unit,

2. each neuron operates in parallel with all other neurons,
3. processing, although simple, characterizes a large number of operations that occur simultaneously, and
4. properties of a neural network exhibit changes that describe a high level of adaptiveness to both new information and new situations.

Constructing an ANN begins by capturing the salient properties of the biological neural network discussed above and creating a formal model. While this procedure may seem straightforward, two important points must be kept in mind. First, a conventional computer is generally a single processor that acts on explicit, programmed instructions. Secondly, the brain is comprised of over 100 billion neurons that function in parallel and enable a single nerve cell to interact with well over 10,000 other neurons (Nelson and Illingworth, 1991; Dayhof, 1990). Neural computing, therefore, is primarily concerned with machines, not brains, and the artificial neural network model is based on its biological counterpart in an extremely general sense. Nevertheless, according to this model, information is processed as a function of the interactions that develop between the processing elements that comprise the network. Knowledge is represented by the relative strength of the interconnections between the processing elements, where each input of knowledge describes a pattern of activity spread across the network. With these critical qualifications noted, the properties of an ANN can be examined and the procedures followed in its design can be described.

Components of an Artificial Neural Network

Although there are many different types of neural network architectures, a typical ANN has three principal features: (1) an organized topology or geometry of interconnected processing elements that form the network's basic structure, (2) a method of encoding information, and (3) a method of recalling information (Simpson, 1990). The characteristics of an ANN that highlight its essential features have been summarized by Rumelhart (1986) and include:

- a set of processing elements,
- a state of activation,
- an output function for each element,
- a pattern of connectivity among elements,
- a propagation rule,
- an activation rule, and
- a learning rule.

In order for an ANN to be precisely defined, each component must be described and the essential components that form artificial neural systems must be addressed. Five central elements or features of an ANN can be identified: (1) the processing elements that form the core of the network, (2) the network topology that explains how the processing elements are arranged, (3) a method of encoding and memory that enables the network to process data and represent knowledge, (4) a mode of learning that defines how the network is trained to generate its response, and (5) a mode of recall that explains how the network produces its response based on what it has learned. Each of these is discussed in turn.

The Processing Element

The processing element may be viewed as the basic building block of an ANN and serves a function analogous to that of the biological neuron. The processing element (PE) is the single component of an ANN where most, if not all, of the computing is performed. A PE computes a weighted sum of its inputs and employs an activation function to transform the calculated weighted sum into an output signal. The fundamental structure of a generic processing element is illustrated in Figure 6.2.

Within this unit, the input signals the PE receives may originate from the environment external to the network or as outputs produced by other processing elements. Whether external or network-derived, these signals form an input vector (X) that can be symbolized as

$$X = \left(\chi_1, \chi_2, \ldots, \chi_n\right), \tag{6.1}$$

where χ_i is the activity level emanating from the ith processing element, or as input to the network. Associated with each connected pair of PE's is an adjustable parameter called a weight. As Simpson (1990) demonstrates, the collection of weights that abut the jth PE, represented as b_j, form a vector as well, which can be written as

$$W_j = \left(\omega_{1j}, \ldots \omega_{ij}, \ldots \omega_{nj}\right), \tag{6.2}$$

where the weight ω_{ij} defines the connection strength from PE a_i to PE b_j. Frequently, an additional parameter exists, θ_j, modulated by the weight ω_{ij}, which

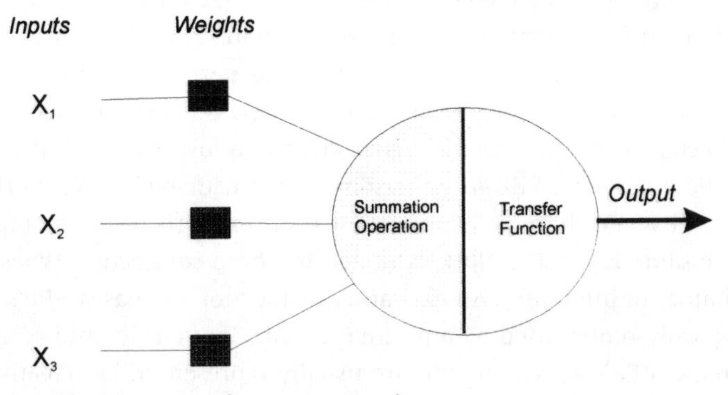

Figure 6.2 Stylized Design of a Processing Element

is associated with the input vector. The term θ is a bias term that is analogous to the threshold level characterized by biological neurons. In an artificial neural network, θ is an internal threshold value that must be exceeded in order for a given PE to be activated.

The weights, their associated values, and the bias term are used to compute an output value given the input signal received. The net input into a processing element is determined by

$$\text{net} = \sum \omega_i \chi_i - \theta. \tag{6.3}$$

Output (b_j) from the processing element is produced by taking the dot product of X and W, subtracting the threshold, and passing the result through a threshold function (f) such that

$$b_j = f\left(\sum \left(\chi_i \omega_{ij} - \omega_{oj} \theta_j\right)\right). \tag{6.4}$$

The threshold or activation function (f) maps a processing element's input to a prespecified range as output. Several threshold functions can be employed. One of the more commonly used is the sigmoid function, given by

$$\text{out} = f\left(net\right) = 1/1 + \varepsilon^{\left(-\text{net}\right)}. \tag{6.5}$$

Network Topology

The topology or architecture of an artificial neural network describes how a collection of processing elements is organized. Network topologies are formed by arranging processing elements into layers and linking the *PE*'s with weighted interconnections. Topologies can therefore be described according to their specific connection type, layer configuration, and interconnection scheme. Many different types of network architectures exist and have been discussed in detail in the neurocomputing literature (Dayhoff, 1990; Hertz, Krogh, and Palmer, 1991).

In general, the architecture of an ANN can be single-layered, bilayered, or multilayered. These fundamental layering schemes can be configured to support one of three principal styles of connectivity: feedforward, feedback, or lateral. Feedforward connectivity occurs when there is a connection from a lower layer to a higher layer. Feedforward signals permit information flow between processing elements in one direction only. Feedback connectivity explains a connection style that links a higher layer to a lower layer, but allows information flow between PE's in either direction (recursively). Network connectivity is lateral when there is a connection from one processing element to another in the same layer. The flow exhibited by these connection types can be either excitatory or inhibitory. An excitatory connection increases a PE's activation and is typically represented by a positive signal. Conversely, inhibitory connections decrease a PE's activation and are usually represented by negative signals.

Encoding and Memory

An artificial neural network does not perform numeric "symbolic" processing as do traditional computer algorithms. Rather, an ANN performs "pattern" processing, which can include any of the following functions:

1. **Pattern Recognition**—where the occurrence of a pattern results in the activation of a PE that is associated with or represents that pattern,
2. **Pattern Production**—where the occurrence of a pattern causes the activation of another associated pattern, or
3. **Pattern Storage**—where the occurrence of an activation pattern changes the network's processing characteristics.

Pattern, as used in this context, can describe a single attribute or a relationship in a data set that takes the form of input to the network model. Examples of patterns that can serve as input to a neural network may include pixel values of characters or graphic elements, digital values representing pictures or other forms of imagery, digital voice patterns, digital signals or measurements from monitoring equipment, coded values from a survey, statistical tables, or field measurements. In each of these cases, the network functions to relate these patterns to a solution that it will produce as output. For instance, a neural network may be developed to simplify data processing and data classification in a demographic application. Such a network may take as its input a vector of population characteristics for a set of observations and produce as output a rating or ranking of those observations based on some quality or condition such as voter behavior, purchasing power, or income category. By taking the input vector of demographic attributes as a pattern, the network attempts to learn the relationship between the demographic variables and the pattern they explain in relation to the desired output solution.

Generally, an artificial neural network can store two types of patterns:

1. **Spatial patterns**—which represent a realization of the real world expressed in the form of a single static image, and
2. **Spatiotemporal patterns**—which explain a sequence of spatial patterns.

Storage of these pattern classes is a function of the network's spatial pattern-matching memory and the hardware environment used to implement the network. Three primary types of spatial pattern-matching memories can be identified: (1) random access memory, (2) content-addressable memory, and (3) associative memory (Simpson, 1990). Mapping a pattern to memory can be either autoassociative or is heteroassociative. An artificial neural network is autoassociative if its memory (M) stores the vectors $X_1, \ldots, X_m$ as a sequence. Heteroassociative memory mapping stores pattern pairs of the form $(X_1B_1), \ldots,$

$(X_m B_m)$. A pattern once stored in memory can be related to a desired output through the recall phase of network processing. However, before a pattern can be recalled from memory, it must be learned by the network.

Learning

Like an infant learning to navigate for the first time, a neural network must learn how to respond to the input patterns it receives from its external environment. Learning, therefore, implies training the network to respond as desired to the stimulus it receives. In an ANN, the set of weight values characterizing its connections explains different states of its memory. These states influence the network's understanding of the input pattern with which it has been presented. As input patterns are presented to the network, the network learns how to respond and ultimately matches the input pattern to the desired response. Stated formally, learning defines any change in memory as the network adjusts or self-organizes to produce an output pattern. This concept has been defined mathematically by Simpson (1990) as

$$\text{Learning} = dW/dt = 0. \tag{6.6}$$

The main problem of learning in an ANN is the process of deriving a function that maps input vectors (stimuli) to output vectors (responses) (Matheus and Hohensee, 1987). To learn this mapping function, the network receives a set of training events (patterns) that it employs to construct an internal representation of the function between input and output. Because learning is a critical phase in developing a neural network, it has received extensive treatment in the literature (Matheus and Hohensee, 1987; Hinton, 1989; White, 1989).

The principal objective of learning is to train the network so that when a set of inputs is presented, the desired output is received. Thus, by sequentially applying input vectors while network weights are modified according to a predetermined procedure, the network learns. During training, the network weights gradually converge to a specified value, which suggests that each input vector produces the desired output with an acceptable level of error or uncertainty. The procedure employed to adjust weights during network training can follow one of several learning patterns, which have been reviewed in detail by Mehra and Wah (1992). Although the list continues to expand, several of the more commonly encountered patterns are:

Correlational Rules
Hopfield Networks
Error-Correcting Rules
The Delta-Learning Rule
Generalized Delta Rule
Cumulative Delta Rule
Hebbian Learning.

Each of the approaches listed above fits a specific type or manner of learning. Four of the more widely known learning modes are:

1. **Heteroassociation**—a mode of learning that involves mapping one set of data to another, and that produces an output pattern different from the input,
2. **Autoassociation**—a learning mode based on storing patterns that reproduce an output pattern that is similar to the input,
3. **Regularity Detection**—a learning mode that searches for meaningful features or associations in data, and
4. **Reinforcement Learning**—a learning mode that describes a supervised form of learning involving feedback.

Implementing a specific learning pattern is generally a function of the category of ANN involved in the application together with the methods built into that network to facilitate its training. Training an artificial neural network to learn can be done by using either a supervised or an unsupervised mode of learning, depending on the network pattern.

If the supervised mode is used, the ANN is presented with a set of inputs for which the desired or appropriate outputs are known. The input pattern and known output response are assembled into training pairs, and the network is trained over a large number of these input-output sets. As the input vectors are presented to the network, an output pattern is calculated, which is then compared to the known (desired) pattern. The difference between the calculated output and the known output defines error. This difference is fed back through the network, and the weights are adjusted according to the selected learning rule. Ideally, changing weights should reduce error as the training set cycles through the network leading to a model that performs as originally conceptualized. As learning continues, errors are calculated and weights modified until the network-produced error converges to an acceptably low level. Backpropagation and Hopfield networks use this approach to learning.

The unsupervised method of learning requires a network to self-organize and generate output that sufficiently categorizes the input. With the unsupervised learning technique, the ANN has no pattern for comparison, and no information is supplied to help determine whether the output response pattern is correct or meaningful. The training set consists entirely of input vectors, leaving the learning algorithm to modify weights to the extent that output vectors are consistent. Therefore, training utilizing this approach relies on the statistical properties of the data used to train the network. Networks that use unsupervised learning include the ART network and the Kohonen self-organizing feature maps.

Once the network has successfully learned the input-to-output mapping, the network represents a specialized source of knowledge. Unlike a conventional computer program where information is a function of its algorithm and the data is stored in memory, or the expert system where knowledge is expressed explicitly as a set of rules coded into a knowledge base, knowledge in a neural network is related to its overall structure expressed implicitly by the state of the network at some equilibrium condition given the pattern it has learned. Therefore, when the network is presented with data, this knowledge, as evidenced by its topology, weights, and mode of interconnectivity, is recalled and applied to the unknown data set. A discussion of the various methods of knowledge recall follows.

Recall

Recall is much like the act of remembering things that have been learned. For an ANN to remember, the patterns stored in memory must be invoked. According to Simpson (1990), if the pattern pairs (A_k, B_k), $k = 1, 2, \ldots, m$, have been stored in memory (Z), a recall function (g) can be defined that takes memory (Z) and stimuli (A_k) as input and returns a response (B_k) as output, or

$$B_k = g\left(A_k, Z\right). \tag{6.7}$$

Two primary recall mechanisms can be noted:

1. **Nearest-Neighbor Recall**—which finds the stored input that most closely matches the stimulus and responds with the corresponding output based on a Hamming or Euclidean distance function, and
2. **Interpolative Recall**—which accepts a stimulus and interpolates from the entire set of stored inputs to produce the corresponding output pattern.

Understanding the components of an artificial neural network is one means of introducing this technology to the decision maker. However, the components and the properties they define are meaningless unless configured into a model that has been designed to address an important decision problem. There are many types of neural networks to select from, a sampling of which are described in the following section, along with their demonstrated utility to environmental decision making.

Neural Network Models

The previous discussion exploring the fundamental elements of artificial neural networks revealed several key concepts that influence both the design and the application of this technology. Those central ideas can be summarized as follows:

- neural networks are not programmed—they learn by example;
- a typical neural network is shown a training set or examples from which the network learns;
- the most common training method employs supervised learning;

- because networks learn by example they have the potential for building computer systems that do not require programming;
- the neural network approach does not require human identification of features or human development of algorithms specific to the problem;
- training the neural network can be time-consuming; and
- neural network architectures encode information in a distributed manner that is shared by many of its processing elements.

With these considerations in mind, specification of the network can begin. Specifying the design of an artificial neural network requires answers to three defining questions:

1. How should the individual processing elements be combined?
2. How many processing elements should be allotted to the input and output layers?
3. How many layers should there be?

While the answers to these three questions are critical to the success of an ANN application, perhaps the most important issue to be resolved in the early stages of network application development is the selection of an appropriate network paradigm (model). Certain network models are better suited than others for solving certain types of problems, and when ANN technology is applied to a problem, paradigm selection takes precedence over all other considerations. In this section, four network paradigms are reviewed with a focus on their fundamentals and their application potential: (1) the Adaline/Madaline network, (2) the backpropagation paradigm, (3) the counterpropagation model, and (4) the self-organizing map. More comprehensive treatments of these and other network models can be found in Dayhoff (1990), Chester (1993), and Lippman (1987).

Adaline/Madaline Networks

The Adaline model employs a threshold logic device that performs a linear summation of inputs and allows weight parameters to be adapted over time. The basic structure of an Adaline unit is illustrated in Figure 6.3. Given this configuration, each interconnection from an input unit has an associated weight value. The derived output value is compared to a known target value, and the difference between these two numbers defines the error term. In an Adaline network, the computed error is used by an adaptation rule to govern and control the process of adjusting weights. The adaline units are binary and can typically explain values between −1 and +1. Output values are scaled in a similar fashion.

Functioning primarily as a summing mechanism, the Adaline unit performs a weighted sum on the input signals according to the relation

$$S = \sum \alpha_i \omega_i, \tag{6.8}$$

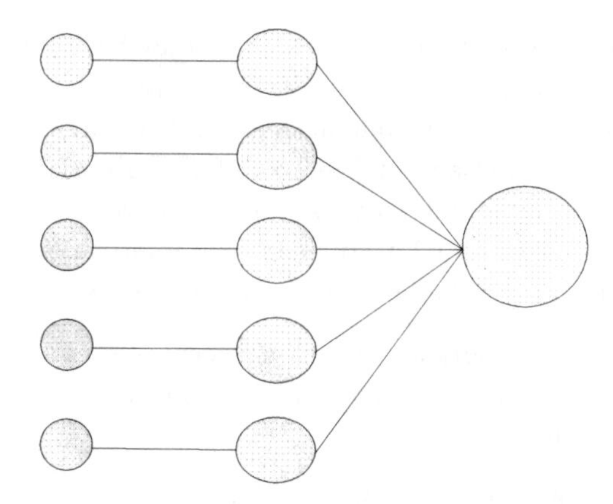

Figure 6.3 Basic Configuration of an Adaline Network

where α represents the output value of the *i*th input and ω defines the connection weight from the *i*th input vector.

Once the weighted sum is calculated, the Adaline unit invokes a threshold function of the general form

$$\text{output } X_i = \begin{cases} +1 & : \text{ if } S \geq \\ -1 & : \text{ if } S < . \end{cases} \tag{6.9}$$

Training an Adaline network is accomplished by presenting a series of input-output pairs to the system. The network learns the relationship(s) between the input-output pairs according to a four-phase process. First, the input set is presented to the Adaline. From here, the Adaline computes its original output. Next, the target output taken from the training set is presented. Then, the adaptation cycle is invoked. During this adaptation phase, the goal is for the Adaline to produce the target output on its own as a function of its learning rule. A common adaptive learning rule applied to Adaline networks is the Widow-Hoff learning rule. The basic form of this rule can be expressed symbolically as

$$\Delta\omega_i = \lambda\alpha i(\tau - X), \tag{6.10}$$

where the change in connection weight, $\Delta\omega_i$, is determined by the values of

λ = a learning constant

α_i = output of unit i

τ = target output

X = output of the Adaline.

The Madaline structure is nearly identical to the Adaline. The main difference between the two models is that the Madaline is simply a multi-layer

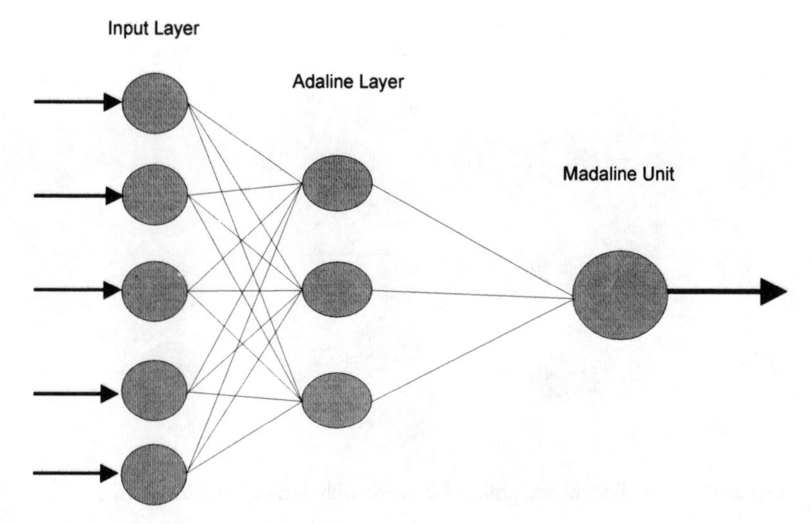

Figure 6.4 Basic Configuration of a Madaline Network

Adaline. Therefore, the Madaline model describes a system of Adaline units connected to form a single Madaline as suggested in Figure 6.4. The layers of a Madaline network are fully interconnected, and weights are defined for the links flowing from the input layer to the Adaline layer. Weights are not associated with the links from the Adaline layer to the Madaline, however.

Training the Madaline network involves presenting a set of input pattern pairs to the system together with their target outputs. Once the input layer is given a pattern, the Madaline unit computes an output that is compared to the target value. Weights are updated following the processing of each pattern pair. As the connection weights are modified, the derived output is compared to the target value. Because in this binary system there either is or is not a match, learning occurs only when the network gives an incorrect answer. When an incorrect answer is noted, the Adaline unit whose sum is closest to zero in the wrong direction is modified. Both Adaline and Madaline networks have found useful application in forecasting, signal processing, pattern recognition, and classification.

The Backpropagation Paradigm

The backpropagation paradigm is one of the most widely used neural network architectures. One reason for its general popularity is that backpropagation is one of the easiest network models to comprehend (Dayhoff, 1990). A typical backpropagation neural network consists of an input layer, and output layer, and at least one intermediate or hidden layer (Figure 6.5). The input layer is composed of one or more processing elements that present raw data to the network, while the output layer is composed of processing elements that store the results of the learned pattern. The hidden layer(s) of the backpropagation

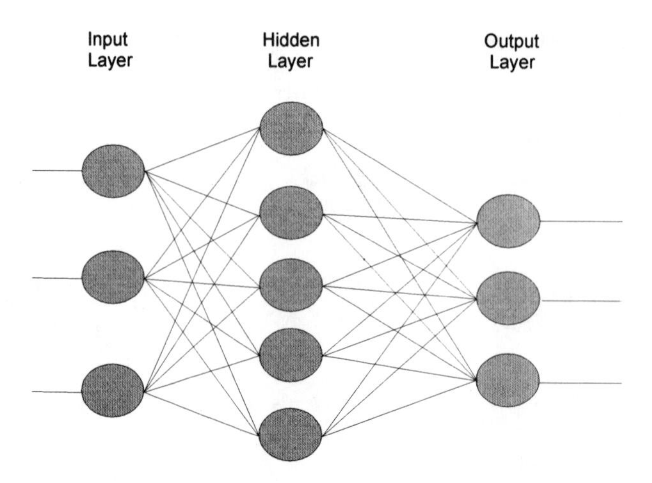

Figure 6.5 A Backpropagation Network with a Single Hidden Layer

network contain processing elements that translate the input data into the desired output information. Consequently, a backpropagation network must learn how to perform this translation.

In the backpropagation paradigm, learning is relatively straightforward. If the network, following the presentation of a training pair, gives the wrong answer, its weights are adjusted as a function of its learning rule so that the error is decreased. Lessening the error term suggests that any future result produced by the network is more likely to be correct. Learning is achieved by presenting examples of the desired output pattern for a sample of input observations. These pattern pairs are shown to the network running in its learning mode, also referred to as forward propagation. In order for the network to learn the relationship between input and output pairs, it must adapt or modify its connection weights in response to the data presented to its input and output layers. Each time the network fails to achieve the desired match, three actions are taken: (1) it adjusts by correcting its weights to reduce the error term, (2) it then cycles through another iteration of the input training examples, and (3) it continues this process until the error term converges to an acceptable minimum value. Wasserman (1989) describes learning as a five-step process consisting of:

1. selecting the training pair from the training set to be applied as the input vector to the network,
2. calculating the output of the network,
3. calculating the error between the network output and the desired output,
4. adjusting the weights of the network in a manner that reduces error, and

5. repeating Steps 1 through 4 for each input vector in the training set until the error for the entire training sample is acceptably low.

The backpropagation network acquires knowledge of the output pattern by altering the power of the existing connections to achieve the desired match. Matching the input signal to the desired output pattern is a function of the learning rule used to train the network. The majority of learning rules applied to backpropagation networks employ external feedback based on the known output values using a supervised training procedure. With a backpropagation network, the updating of activation values or weights propagates forward from the input layer through each hidden layer. As the network modifies its internal weights, the correction process begins with the processing elements of the output layer, then propagates backward through each hidden layer until it reaches the input layer (Kohonen, 1988). Symbolically, a processing element in a backpropagation network will transfer its input according to the following general relation:

$$X_i^{(s)} = f\left(\sum\left(\omega_{ij} * X_i^{(s-1)}\right)\right), \tag{6.11}$$

where f typically represents a sigmoid function defined by

$$f(z) = \left(1.0 + e^{-2}\right)^{-1}, \tag{6.12}$$

although in some cases sine functions or hyperbolic tangent functions are used.

Error in a backpropagation network can be expressed as

$$E_j^{(s)} = f\left(\sum\left(\omega_{ij} * X^{(s-1)}\right) * \sum\left(E_k^{(s+1)} * \omega_{kj}^{(s+1)}\right)\right) \tag{6.13}$$

and is reduced by modifying the value of the weights through a gradient descent rule of the general form

$$\Delta\omega_{ij} = Lcoef * \left(dE/d\omega_{ij}^{(s)}\right). \tag{6.14}$$

Calculating the partial derivative in Equation 6.14 yields the expression

$$\Delta\omega_{ij} = Lcoef * e_j^{(s)} * X_i^{(s-1)}, \tag{6.15}$$

where *Lcoef* defines the value of a predetermined learning coefficient used to train the network, and E explains the global error described across the layers.

While backpropagation networks have enjoyed applications in a wide variety of implementations, they characterize a long and uncertain training process that can be fraught with difficulty (Wasserman, 1989). Nevertheless, this paradigm remains a powerful model with a high degree of general pattern-matching capabilities.

Counterpropagation Networks

Counterpropagation networks, while not as readily generalized as back-propagation models, provide a solution for applications that cannot tolerate long training sessions. The counterpropagation paradigm is a combination of two well-known algorithms: Kohonen's self-organizing map, and the Grossberg Outstar (Wasserman, 1989). In its most fundamental implementation, a counterpropagation network functions as a lookup table capable of generalization. The network selects from a set of examples by allowing them to compete with one another. Then, through a combination of normalized inputs and competition between inputs, the nearest-neighbor association between input and output provides the rationale for determining a match. To achieve an input-to-output match, a training process associates an input vector with its corresponding output vector. Once trained, the application of an input vector produces the desired output vector, and, because of the network's capacity for generalization, the correct output can be produced even in situations where a given input vector is partially incomplete or partially incorrect.

In the design of a counterpropagation network, the Kohonen layer acts as a nearest-neighbor classifier with the input layer serving as a buffer. The processing elements in the Kohonen layer compete as the normalized input signals are received. Competition describes the condition where the node with the highest output value wins and produces a logical one as output, while the remaining nodes produce values of zero. Associated with each node in the Kohonen layer is a set of weights that connect each node to each input. Weights are adjusted so that similar input vectors activate the same node. This feature enables the Kohonen layer to classify the input vectors into groups that are more or less homogeneous. However, because the Kohonen layer is trained using an unsupervised method of learning, it is difficult to predict which node in the Kohonen layer will be activated.

The output layer of a counterpropagation network uses a Grossberg Outstar (Wasserman, 1989). A Grossberg Outstar is basically a processing element that learns to produce a certain output signal only when a specific input signal is applied. Since the Kohonen layer produces only a single output, the Grossberg Outstar provides a means of decoding that input into a meaningful information class. Decoding is accomplished by taking the weighted sum of the Kohonen layer outputs to produce a net output for each node in the Grossberg layer. This output is derived by the following relation:

$$\text{net}_j = \sum k_i \omega_{ij}. \qquad (6.16)$$

Training the Grossberg layer to yield this translation is a four-step procedure:

1. apply the net input vectors,
2. establish the Kohonen outputs,

3. calculate the Grossberg outputs, and
4. adjust weights if it connects to a Kohonen node having a nonzero output.

It should be noted, however, that Grossberg training is supervised. This implies that the algorithm has a desired output that is used to direct its learning. The unsupervised, self-organizing operation of the Kohonen layer produces outputs at indeterminate positions in the logic of a counterpropagation model. These signals are then mapped to the desired outputs of the Grossberg layer. A more detailed explanation of counterpropagation networks can be found in Wasserman (1989).

The Kohonen Self-Organizing Feature Map

Kohonen's self-organizing feature map defines a two-layered network consisting of an input layer set to the number of variables characterizing the input pattern, and a second, competitive layer (the Kohonen layer), expressed as a two-dimensional grid (Figure 6.6). Links in this network flow from the first layer to the second, with the two layers fully interconnected. When presented with an input pattern, each node in the first layer assumes the value of the corresponding entry in the input pattern. The nodes in the second layer sum their inputs and compete to find a single winning unit that best reflects the input signal. Competition, in this context, describes the process where the nodes in the second layer measure the Euclidean distance of their weights to the incoming input value, such that

$$D_i = \left(x_1 - w_{i1}\right)^2 + \left(x_2 - w_{i2}\right)^2 + \ldots + \left(X_m - w_{im}\right)^2, \tag{6.17}$$

where D_i is the computed distance of each of the n-nodes in the second layer, x is the input value, and ω is the weight.

During recall, the processing element in the second layer with the minimum distance is the "winner" and transmits an output value of 1.0, while the remaining nodes send out a value of 0.0. The logic underlying the signaling process is

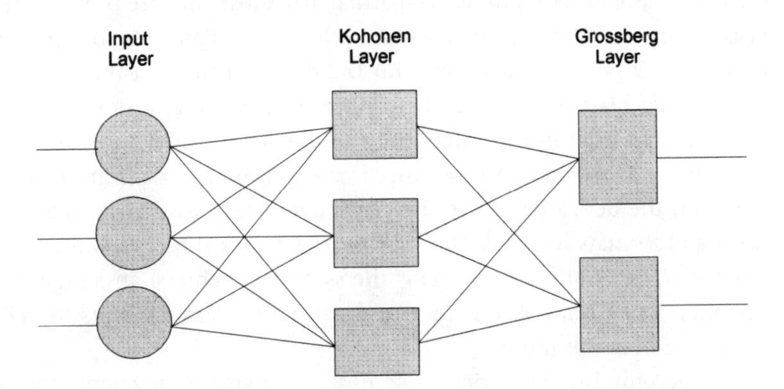

Figure 6.6 A Generalized Feedforward Counterpropagation Network

fairly simple: the winning node, as measured by its Euclidean distance, is the closest one to the input value and is therefore assumed to represent that input value. This logic suggests that the input data, which can have many dimensions, is represented by a two-dimensional vector that can preserve the order of input data of higher dimension. A detailed explanation of this signaling logic and the basic operation of self-organizing networks can be found in Kohonen (1988).

The initial step in the operation of a Kohonen network is essentially to compute a matching value for each node in the competitive layer. This value expresses the degree to which the weights of each processing element match the corresponding values of the input pattern. Once this winning node is identified, the neighborhood around that processing element must by delimited. The neighborhood, as described by Kohonen (1988), defines those processing elements in the second layer that are closest to the winner. Once this neighborhood has been defined, the weights are updated for all the nodes that form the neighborhood. This adjustment assumes that the winning node and its neighbors are becoming more similar to the input pattern than are the remaining processing elements of the second layer. Then, as the network learns, the size of the neighborhood decreases iteratively and connection weights are modified to incrementally improve the solution. Training a Kohonen network in this manner relies on the method of unsupervised learning. Once trained, Kohonen networks have found useful applications in problems involving predictions and data categorization where data is noisy and the underlying structure of the input is not clearly understood.

Developing Neural Network Applications

Medsker and Liebowitz (1994) have likened the process of developing an artificial neural network to that of designing an expert system. But while there are certain similarities between the two, there are also key differences. In terms of similarities, both technologies require their builders to acquire and represent a body of knowledge as a set of facts, rules, or data. However, unlike an expert system, neural network models—particularly those that are trained using supervised learning—demand the accurate selection of data sets to represent examples of the input data together with the desired result. In the case of network models that learn by means of unsupervised training, a domain expert must examine and interpret the network's output to consider the effectiveness and correctness of the network-produced categorizations. With an artificial neural network, the developer must also place data sets into their proper representations and formats for presentation to the network during training. Finally, the builder of a neural network trains the system to achieve an acceptable level of accuracy. Once trained, test cases are run with known results to validate and verify the network model.

To accomplish the varied tasks outlined above, a development methodology or strategy is needed. Although the development of a neural network application can be described in a manner similar in concept to the structured design methodologies applied to traditional information systems, certain aspects of the process are unique to neural network technology and demand more detailed consideration.

A relatively straightforward method for developing a neural network was introduced by Bailey and Thompson (1990) and consists of four phases: (1) a concept phase, (2) a design phase, (3) an implementation phase, and (4) a maintenance phase. The concept phase is largely concerned with selecting the application, bounding the problem, and determining the basic functionality of the system. The concept phase also involves selecting the appropriate network paradigm, given the nature of the problem. In principle, the properties of the problem should be matched directly with the features and functional considerations of a given paradigm (Medsker and Liebowitz, 1994). If training data is available, a supervised learning method may be employed and a network model that learns via supervised training can be selected. However, in situations where a simple categorization of the data is not possible, an unsupervised approach may be required and the appropriate network model will be one that employs unsupervised learning. The design phase uses the specifications formulated during the concept phase to develop a detailed design of the system. This design phase surrounds the issues of developing the network according to the selected network paradigm and includes decisions as to the number of processing elements, layers, initial weights, learning rules, and other parameters necessary to create the physical network. The implementation phase centers around establishing the network environment, realizing the design as computer code, and training the network model. The maintenance phase follows implementation and is concerned primarily with evaluating system performance and modifying the design to improve performance if required.

An alternative approach to network development was offered by Turban (1992). This method, illustrated in Figure 6.7, is a nine-step procedure, but we have summarized it here into six main tasks:

1. collection of training and testing data,
2. specification of a network structure and learning algorithm,
3. initialization of critical network parameters,
4. transformation of inputs into a network-compatible format,
5. network training and evaluation, and
6. implementation of the application-ready model.

Regardless of which general approach is followed when developing a neural network, several critical issues dominate the process. These include application selection, paradigm selection, system configuration, and network training and testing (Medsker and Liebowitz, 1994). Because of their importance to the success of neural network application, each of these considerations is discussed in the following sections.

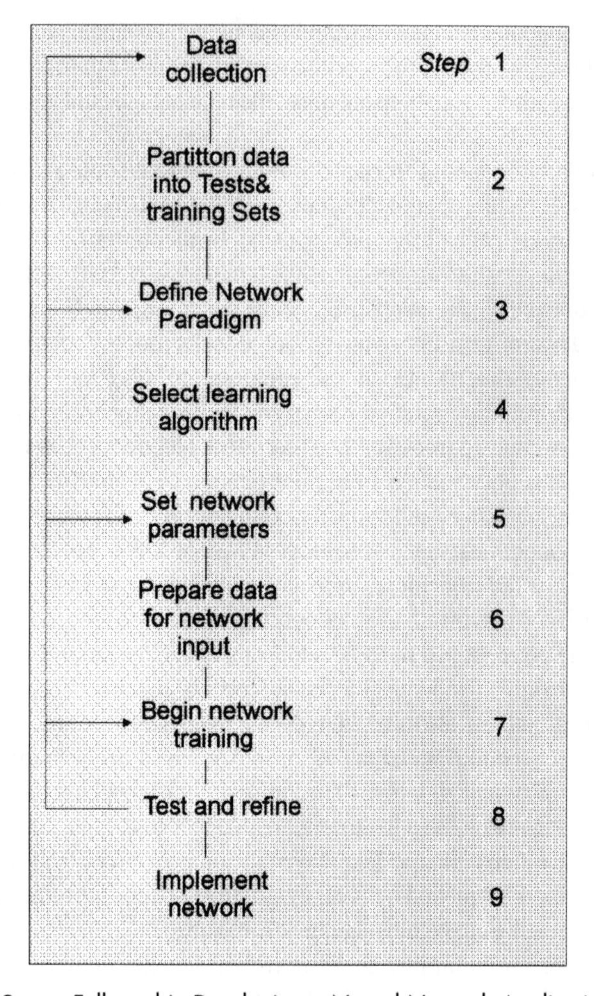

Figure 6.7 Stages Followed in Developing a Neural Network Application

Application Selection

To make any information technology work, it must find the place in the decision-making process that it has been targeted to support. The artificial neural network is no exception to this rule. In general, artificial neural networks are well suited to decision problems where:

1. standard technology is inadequate, ineffective, or too difficult to implement,
2. qualitative or complex quantitative reasoning is involved,
3. many highly interdependent parameters are involved, rendering a clear understanding of the analysis and solution of a problem difficult,

4. data is intrinsically noisy or error-prone, and
5. project development time is limited, though the time required to train a neural network is considered reasonable.

Paradigm Selection

The number of neural network models to select from is large and many new models are under development. When selecting a network architecture and learning method, consideration must be given to four network-specific architectural features: (1) the size of the network, (2) the nature of the inputs and outputs, (3) the memory mechanisms employed, and (4) the type of training method used. First, the application problem must be carefully evaluated to determine whether the number of nodes required to represent the input and produce useful output is feasible given the constraints of the available computer installation. Next, the memory mechanism employed by the network model must be examined since it will influence how the network recalls patterns, particularly when partial information is involved. Finally, the required method of training characterizing the prospective network model must be assessed relative to the type of data available and the time constraints involved in teaching the network.

Application Design

Designing the application involves establishing the specifics surrounding network topology and network parameterization. The critical decisions facing the developer with respect to design include determination of the number and types of processing elements needed at the input and output layers, the number and size of the hidden layers (if any), and the general pattern of connectivity that will be used. Once these questions have been resolved, a decision must be made regarding which learning algorithm and transfer function to use. Because network paradigms differ in terms of design specifics, the options available to the developer will vary.

In addition to the requirements imposed when configuring the network's topology, application design also directs attention to the data collection issues that influence creation of the network training and testing data sets. Data collection to produce representative and sufficiently general training and testing cases must be based on a complete analysis of the application problem. This will insure that the problem is well bounded and that the context of the neural network is well understood. With respect to data collection, training and testing data should fulfill three conditions:

1. it should be representative of the problem,
2. it should cover the widest scope of the problem that is practical, and
3. it should include extreme cases or exceptions and conditions at the boundaries of the problem.

Although the more data available to train the network, the better, it should be noted that larger data sets tend to increase processing time during training.

System Configuration

Once the design specifics of the network have been decided, several parameters must be selected in order to facilitate network learning. While the parameters involved will differ depending on the network paradigm, the most common values that must be initially supplied by the developer include the rate of learning, learning coefficients, processing elements or layer weights, threshold values, and error convergence terms. Presently, very little guidance is available in the selection of these terms. Therefore, a degree of trial and error can be expected before the appropriate values or parameters are set. The trial-and-error aspects of developing an application manifest themselves during the training and testing of the design.

Training and Testing

During training, the initial network configuration is subjected to a rigorous and critical evaluation of its performance. Training, as applied to the ANN, is an iterative process that involves presenting the training data set to the network. As the training data is shown, the network strives to adjust its weights and produce the desired output for each input with a minimum of error. With the presentation of each input vector, weights are modified until a consistent set is derived that works for all observations that form the training data set. Because the choice of the network's structure coupled with the parameters selected for network initialization conspire to determine the time required for network training, careful consideration should be given to these factors early in the development process.

The concept of training a network can be somewhat misleading. In practical experience, any network can be trained, particularly if training is simply defined as a model that converges to a user-determined level of error. A network that converges and ceases training, however, does not imply that the network has learned. The degree to which the network has learned to relate input patterns to the desired output pattern cannot be determined until it has been tested. In many instances, an artificial neural network may not be expected to perform perfectly, but a certain level of performance is always required.

The network testing process examines performance using the weights derived through training by measuring the ability of the network to classify (process) the test cases correctly. Testing, therefore, requires developing a test plan that includes routine examples as well as extreme cases. If testing reveals large deviations between what the network produces as output and the known values or conditions of the test data, corrective measures such as the following may be required:

1. reactivating the training process,
2. reexamining the quality of the training data,
3. modifying the configuration of the network, or
4. reconsidering the selection of a network paradigm.

Because the weights and connections derived through training are difficult, if not impossible, to interpret, the neural network is typically viewed as a black box during testing. From this perspective, tests are performed on the relationships between the inputs and outputs, and between the network-produced outputs and their known categories. In the majority of cases, the "correctness" of the network, when compared to the test data, is determined by means of statistical testing.

Three general statistical procedures can be employed to test the network's performance. A network designed for predictive purposes can be effectively evaluated using correlational testing or regression analysis. Correlational or regression tests help establish the relative degree of association or statistical "fit" between the network-derived output and their known values. For a network created to perform classification or pattern recognition, X-square, or contingency table analysis employing a confusion matrix can be employed. Since the majority of neural network applications to date involve some aspect of data classification, accuracy assessments based on contingency table analysis provide the most useful insight regarding network performance. By means of contingency table analysis, an error matrix is produced that effectively summarizes errors of omission and commission and provides estimates of the network's accuracy and error rates. Regardless of the technique employed, statistical testing yields a summary measurement that communicates how well the network performs according to its designed task. Based on an assessment of this value, the decision can be made to either retrain the network to improve its performance or implement the model as a decision-assisting device.

Neuro-Decision Making

Automation is an essential trend in decision making. For the decision maker faced with the need to process large volumes of information quickly, see patterns and relationships in complex or noisy data, categorize situations and make appropriate choices, automation is a necessity and perhaps the only effective means of simplifying the decision process. The need for simplification is particularly evident when a disproportionate amount of the decision maker's time and effort involves processing routine information, or when time and expense are expended on complex issues with uncertain outcomes or limited payoffs. In situations such as these, the artificial neural network can make an invaluable contribution to the decision-making process.

An ANN has demonstrated utility in several functional areas germane to decision making: pattern recognition, learning, classification, generalization, and abstraction. In addition, the neural network can help interpret incomplete or noisy data as well. Artificial neural networks also have the potential to provide some of the human problem-solving characteristics that cannot be efficiently simulated using the logical and analytical techniques of expert systems, or by means of standard software technologies (Turban, 1992). Here, the examples cited include the ability of an ANN to analyze large quantities of data to establish patterns and characteristics in situations where rules are not known, or when confronted with inherently noisy or incomplete data. In these cases, the ANN can provide the human characteristics of interpretation and abstraction in an attempt to make sense out of the data. Several other important factors linking ANN technology to decision making include:

Fault Tolerance—because an ANN has many processing elements, damage to a few will not bring its processing capabilities to a halt.
Generalization—when presented with previously unseen, incomplete, or noisy data, the network generates a reasonable response as output.
Adaptability—the network is capable of learning in new environments.

These three aspects of ANN technology alone suggest that neurocomputing defines problem-solving characteristics that differentiate neurocomputing from traditional computing approaches—characteristics that can be effectively exploited by the application developer.

Development of neural network applications should therefore concentrate on problems where data is multivariate and describes a high degree of interdependence between attributes, or on situations involving many hypotheses that are to be pursued in parallel and that require high rates of computation. It is also possible to combine neurocomputing with existing information technologies to produce more powerful hybrid systems. Examples of such integrated systems include coupling ANN technology with geographic information systems, expert systems, and decision support systems. Through this form of integration, information technologies are created that better mimic the human decision-making process, even under conditions of uncertainty, and provide solutions to a wider range of complex decision problems. In the following section, examples of neural network applications directed at environmental decision making are examined along with several attempts to describe the integration of ANNs with existing information technologies.

Environmental Applications

Neurocomputing provides a new approach to solving existing problems and addressing new ones. From the standpoint of an information technology, the appeal of neurocomputing is its ability to yield simplified descriptive explanations from complex multivariate databases. Thus, by means of an artificial neural network, the decision maker is offered a tool that can see patterns in data and explore anomalies that present techniques cannot reveal. Current applications of neural network technology focused on problems related to environmental analysis and environmental problem solving fall within four general categories:

1. image classification and interpretation,
2. environmental prediction and simulation,
3. landscape analysis and mapping, and
4. data description and modeling.

The concepts, goals, and approaches that guide artificial neural network development in each of these application areas are reviewed below, beginning with the problem of image classification.

Image Classification

Collecting information concerning Earth's vital resources has been a fundamental need and a continuing problem within the earth and environmental sciences. The science of remote sensing has done much to provide data collection techniques and instruments that have greatly expanded the analyst's ability to measure and map critical earth surface phenomena. It has also contributed to a form of information overload, where large quantities of data can be collected and acquired for any given location on Earth's surface, through rapid, cost-effective methods of processing data that are not widely available. Here, the artificial neural network, functioning as a trained image classifier, has shown promise as a possible solution to the problem of image processing (Jensen, 1995).

Utilizing remotely sensed data directs attention toward the analysis and classification of images that depict properties of Earth's surface as values of reflected or emitted energy. Through the analysis of these images, data is classified into information products that take the form of thematic maps (maps that display a specific theme such as forest cover, land use, or crop type). Traditionally, translating the raw values of reflectance or emittance recorded in digital form on the image and placing those values into meaningful informational categories has relied on the use of statistically based algorithms. These algorithms employ one of several approaches to cluster numerical patterns into groups that are more or less homogeneous and can be equated with the nominal classes of interest that will form the thematic map.

A major limitation of the statistically based method of image classification is its reliance on human guidance and expertise to (1) select representative statistics to implement the classification procedure, (2) select the appropriate classification algorithm, and (3) apply the algorithm to the data and interpret the resulting thematic map. Where that expertise is unavailable or in short supply, the potential benefits that could be derived from the use of remotely sensed data cannot be realized. Adopting a neural network approach to image classification replaces or augments the human expert, simplifying Steps 1 through 3 above. By drawing on the pattern recognition capabilities of the neural network, an "intelligent" classifier can be produced that is capable of performing the knowledge-intensive aspects of image classification without human guidance. The ANN, therefore, becomes a knowledge-specific model that can be created to achieve the goal of transforming raw digital data into information categories. Recent research on the application of neural networks in image classification has shown that network models can achieve accuracies equal to or exceeding those obtained via statistically based methods and are capable of generalizing beyond the data used in their testing (Benediktsson et al., 1990; Hepner et al., 1990; Heermann and Khanzeni, 1992; Dreyer, 1993; Civco, 1993; Foody, 1995; Gong, 1996).

Today's neural networks developed for image classification are constructed using the backpropagation paradigm. Because the backpropagation model learns based on known categories, the training and testing of the network model can be better controlled and its results more clearly understood. Demonstrating the utility of backpropagation networks, Hepner et al. (1990) created a fully interconnected model consisting of three layers designed to produce land cover classifications using data acquired from a Landsat Thematic Mapper scene. The first layer of this network was composed of a 3 × 3 × 4 array of processing elements, used for processing the input data. This configuration of the input layer permitted a 3 × 3 pixel window to move across the four Thematic Mapper bands that were selected for use in this study, assuring a simultaneous consideration of texture as well as spectral decision-space parameters. The second layer was a single, ten processing element, hidden layer, while the output layer was composed of four processing elements representing the four target classes of land cover (water, grass, forest, and urban) to be produced by the network (Hepner et al., 1990). Preliminary assessment of this configuration showed that network-produced classifications transcend spectral variance and perform well when given only the purest and most minimal of training classes.

In a similar study, Civco (1993) constructed a three-layer backpropagation network to identify seven categories of land use/land cover using Landsat Thematic Mapper imagery. When applied to their respective test data sets, the network models described in this paper produced land cover maps with classification accuracies ranging from 34 percent for coniferous forest cover to 99 percent for barren land. Finally, Dyer (1993) tested the appropriateness of the backpropagation paradigm for land cover classification using data acquired from the SPOT satellite system. The network, describing a model comprised of two hidden layers, was applied to test data sets of contrasting segment size. Based on the test results obtained, network-produced classifications defined accuracies in the range of 70 to 80 percent, with optimal performance obtained via a classifier that considered only spectral signature as primary input.

Although it may be premature to generalize on these early studies, it appears that a standard backpropagation network, consisting of an input layer, one or two hidden layers, and an output layer, that employs a sigmoid transfer function and some variation of the delta learning rule, provides that most useful topology of image classification. When such a model is compared to existing statistical methods, it generally tends to produce similar or higher accuracies and has the additional advantage of being distribution-free. However, models of this variety can be computationally complex and slow to train.

Environmental Prediction and Simulation

Simulating the behavior of environmental parameters and forecasting changes in the state of key environmental variables are activities central to the needs of the environmental decision maker. Although a wide assortment of computer models have been developed to characterize environmental process and predict the status of selected environmental systems and variables, many environmental processes and the variables that define them are nonlinear and not easily described by physically based equations. The role of an artificial neural network as a modeling and simulation tool has been demonstrated recently by Cook and Wolf (1991), French (1992), Boznar (1993), Derr and Slutz (1994), Dowd (1994), McGinnis (1994), Gimblett and Ball (1995), and Pattie and Haas (1996).

When applied in a modeling or simulation context, the pattern-processing capabilities of an ANN are called into play and generally involve some form of pattern production, where the input data received by the network causes activation of an associated pattern that the network has learned. For example, Cook and Wolfe (1991) developed a backpropagation network to predict monthly average temperature based on the patterns displayed by the input data. The input data presented to the network consists of a pattern of mean daily maximum temperature for the current month and the previous month, precipitation, relative humidity, and minimum temperature of the current month. Using this input, the model maps the pattern revealed by the data to a pattern of temperature that can be carried out over three months into the future.

In a somewhat related example, Derr and Slutz (1994) describe a backpropagation model that was created to predict sea surface temperature and the onset of an El Nino event. This network model, composed of twelve input nodes, each representing the pattern of a selected climatological variable, a hidden layer of thirty nodes, and a single output node, provides forecasts that have an average RMS value of 0.76 °C deviation from the actual temperature. Moving to a different type of predictive problem, Boznar (1993) describes a neural network created to predict short-term values of sulfur dioxide concentrations. The model, designed to assist in power-plant plume analysis, performs continuous monitoring of carbon dioxide concentrations and was shown to produce results superior to those obtained through conventional modeling techniques.

Landscape Analysis

Evaluating land resources and allocating functions to their appropriate spatial locations place a premium on the decision maker's ability to (1) process and synthesize multivariate data sets and (2) derive an expression of land quality or suitability to guide and support the decision-making process. Recognizing the inherent complexity of land resource planning, Yin and Xu (1991) introduced a neural network model designed to perform multi-objective land use analysis. The network, following the backpropagation paradigm, describes an input layer consisting of twenty-two nodes, each representing the pattern of a critical landscape factor that exerts a recognized influence on land use options. The data items are arranged in a database format where each landscape factor in the database has associated with it certain "developmental" attributes such as yields, net economic return, soil erosion rating, and habitat value. This is the data presented as input to the network. The network model also describes a hidden layer of four nodes that act as a feature extractor and function to transform the raw input patterns into higher-level features that are more effective at achieving the nonlinear mapping between the landscape variables used as input and the desired output pattern. The output layer of this model consists of two nodes designed to yield a binary (0 or 1) response to a given pattern of input. The value 0 or 1 indicates the appropriateness of a specific land use option under consideration (Yin and Xu, 1991).

In a related application, Wang (1994) describes a three-layer backpropagation model designed to address the problem of agricultural land suitability assessment. A major feature of this study was that the networks developed to perform land suitability assessment for different crop types were integrated with a GIS by means of two computer routines that manage data communication from the GIS to the ANN and from the ANN to the GIS (Wang, 1994). Although the two software systems stand separately, the integration software enables the two systems to interact, allowing the user to select data items from the GIS and implement the desired ANN model, whose results are then passed back to the GIS for display.

The question of environmental carrying capacity assessment and mapping was explored by Lein (1995) using an artificial neural network. In this study, a three-layer backpropagation model was developed to translate Normalized Difference Vegetation Index (NDVI) measurements obtained from the National Oceanic and Atmospheric Administration (NOAA) Advanced Very High Resolution Radiometer (AVHRR) series satellites into estimates of population support capacity. The rationale for using NDVI measurements as an indicator of population support potential was based in part on the assumption that a relationship exists between the physical structure of an ecosystem and its capacity to provide resources to support human populations at varying levels of technology. NDVI measurements were also selected in this application because of their well-recognized capabilities for capturing ecosystem response to deforestation, drought, and other biophysical processes (Lein, 1995). Using both the pattern-recognition and pattern-processing abilities of an ANN, the network developed for this study achieved a 77.5 percent level of accuracy in comparison testing against a reference data set for the study area in equatorial Africa and illustrated the deductive power of ANN technology when applied to the problem of environmental monitoring.

Data Description

In the course of environmental decision making, situations frequently arise where data contains unusual characteristics that require careful interpretation. In these instances, illuminating the underlying pattern or structure of a data set can provide useful insight to the decision maker. However, because there is no known or desired output available to train a network, paradigms such as back-propagation that require supervised training cannot be applied. In a study conducted by Winter and Hewitson (1994), the Kohonen Self-Organizing Feature Map (SOM) paradigm was used to organize demographic data gathered from census information. Through the use of the SOM model, the researchers were able to investigate population groupings and their geographic distribution. The SOM network, applied in a nonlinear, unsupervised manner, demonstrated a unique ability to identify important social structures and demographic patterns in the data set. However, because the network output is the product of an unsupervised method of training, correctly interpreting the results obtained from the model, particularly those patterns that were more subtle in character, was difficult and suggests the need for a refinement of this technique (Winter and Hewitson, 1994; Openshaw, 1994).

Neural Network Development Tools

The artificial neural networks discussed in this chapter are software applications that require coding into a machine language for execution on a digital computer. Two basic approaches exist for programming ANN applications, and each approach parallels, in concept, the development tools and environments available for building expert systems. For the decision maker interested in developing neural network applications or testing the feasibility of neural networks for a specific problem task, artificial neural networks may be coded using a standard programming language or a specialized neural network development tool. Although programming languages such as C or Fortran have been used to program the various network paradigms described in this chapter, from a developer's perspective, programming languages are somewhat rigid and lack the ease of implementation and modification found in network development tools. For these reasons, coupled with the programming complexities involved in writing ANN code, development tools are an attractive alternative.

Artificial neural network development tools are software packages in which the standard computations required for creating the neural network have been preprogrammed. Some development tools are similar to expert system shells, describing software applications where the major features of a selected network paradigm have been provided and only require selection of a learning algorithm, transfer and summation function, and connection weights. Other development tools are more comprehensive in scope and broader in functionality. These systems provide capabilities to create any one of several network paradigms along with the mechanisms needed to select specific network design options. In the majority of systems available, network development tools support a wide assortment of network models and contain specialized features that permit:

1. editing and modification of the initial network configuration,
2. dynamic selection of weights, learning algorithms, and related network parameters,
3. real-time visualization of the learning process via histograms and other on-screen displays,
4. options for creating run-time or compiled versions of the created network model, and
5. data sharing with commercial spreadsheet and database management packages.

An extremely useful guide to neural computing development tools detailing the features of a large selection of commercial systems was published in *AI Expert* (July, 1991).

Although network development tools greatly simplify the stages followed when creating a testable neural network model, the network developer still needs to perform several preprocessing and postprocessing tasks including:

1. programming the format of the database that will be presented to the network,
2. partitioning the data into training and testing files,
3. transferring the data into file structures suitable for input to the neural network, and
4. transferring the output generated by the network into file formats suitable for secondary applications.

Neural Networks and Uncertainty

The inherent error and uncertainty of the input vectors presented to a neural network also become features of the network-produced output. The presence of error has encouraged development of network models capable of reasoning with imprecise and inexact data. One class of network models introduced for this purpose is based on fuzzy modeling concepts (Horikawa et al., 1992; Pal and Mitra, 1992; Takagi et al., 1992; Keller et al., 1992). Indeed, an increasing number of researchers have been exploring the subject of fuzzy neural networks as a means of creating more powerful tools for processing fuzzy information (Gupta and Qi, 1992). Although a fuzzy neural network is designed to function similarly to the nonfuzzy model, the fuzzy neuron has the ability to cope with information that is inexact or imprecise. The main difference in a fuzzy neural network is that the inputs to the fuzzy neuron are fuzzy sets. According to Gupta and Qi (1992), these fuzzy sets may be labeled by linguistic terms such as high, large, or warm, or by other representations of data that are inherently imprecise. These inputs are then weighted following a logic that differs from that used in a nonfuzzy neural network. The weighted inputs are then aggregated, not by a summation function as would normally be the case, but by a fuzzy aggregation operation. Several fuzzy neuron models are described at length by Gupta and Qi (1992).

The rationale for developing fuzzy neural networks is to enable these systems to approach real-life situations in a manner more like humans do (Pal and Mitra). To that end, fuzzy concepts have been incorporated into neural networks designed to model output possibility distributions, learn and extrapolate complex relationships between the antecedents and consequents of rules, and apply fuzzy reasoning (Pal and Mitra). To the environmental decision maker, two application areas appear to have particular merit. One is the development of fuzzy neural networks capable of performing fuzzy classification of patterns contained in a data set. The other is the design of neural network models that can perform fuzzy logic inference.

The problem of fuzzy logic inference has been addressed by Keller et al. (1992). In this paper, researchers present a fixed network architecture based on an evidence aggregation paradigm that employs the fuzzy union and fuzzy intersection operators. In a series of experimental tests used to model several linguistic concepts, the network was shown to possess the desirable theoretical properties of a fuzzy inference mechanism and produced reliable inference results. With respect to fuzzy classification, Pal and Mitra (1992) describe a fuzzy neural network model based on the multilayer perception using the backpropagation paradigm. This network, taking as input a vector of membership values to linguistic properties, produced an output vector defined in terms of fuzzy class membership. When applied to the problem of speech recognition, the network demonstrated the capacity to efficiently model fuzzy or uncertain patterns in a manner comparable to Bayesian classifiers.

The artificial neural network, however, is merely one example of an emerging technology that can find useful application on the environmental decision maker's desktop computer. Beyond its role as an alternative to traditional algorithmic approaches to data analysis, neural computing can be combined with conventional software systems to produce hybrid technologies with the potential to create computer systems that can mimic human decision making even under conditions of uncertainty. All that remains is to better define the application potential of these systems and critically evaluate their expanded role as decision support tools. In Chapter 7, the potential of neural networks and the other information technologies described in this book are reexamined with a critical eye directed toward tomorrow.

7 Tomorrow's World

We live in a rapidly changing world. The programs and policies that shape environmental decisions today are the blueprints for tomorrow's world. When a decision is made today, it is essential to ensure that it will be appropriate in the future as well. The future is not just an abstract, distant point in time, but a real place that will be populated by living things. It represents a steady evolution, punctuated by occasional discontinuities, where any situation viewed along the time continuum will be subject to constant forces of stress and adjustment. Consciously or unconsciously, the environmental professional adopts this view of the future whenever a decision is contemplated. This almost intuitive process sometimes results in a "good" decision, but not always. In either case, the quality of an environmental decision can be attributed ultimately to the judgment, insight, and creativity of the decision maker.

Environmental decision making is a type of forecasting that tries to create a better tomorrow by systematically converting data into information that can support the decision process. It is predicated on the thesis that the application of technical judgment to comprehensive data, collected systematically and analyzed rigorously, will produce sounder results than simple intuition alone will produce. In order for this thesis to be accepted, three issues must be resolved:

1. information availability,
2. methodological practicality, and
3. uncertainty.

Of course, all environmental decisions are subject to uncertainty due to several factors that are difficult to control, including:

- the accuracy of data relating to environmental conditions,
- the inherent uncertainty of the processes involved,
- the error surrounding the methods employed in analysis, and
- the limitations in knowledge and judgment regarding the environmental.

As a result, it will never be possible to explain the future with absolute certainty or to produce a perfect environmental decision. However, approaches should be explored to reduce the impact of uncertainty. Increasingly, these approaches capitalize on the computational and data storage capabilities of digital computers, and collectively they define a set of information technologies that complement the decision maker and direct the decision-making process. Those technologies include the geographic information system, the decision support system, the expert system, and the artificial neural network. These information technologies were reviewed earlier in this book, and their fundamental characteristics were discussed relative to the needs and issues that motivate environmental decision making.

Beginning with a treatment of the environmental decision-making process and the cognitive theories and pragmatic issues that direct decision making, a foundation was constructed in Chapter 2 that was built upon as each technology was examined. In Chapter 3, the concept of a geographic information system was described and its role in decision making was defined. Chapter 4 introduced the decision support system as a technology that blends the database environment with modeling and simulation approaches that can explore ill-structured problems. Artificial intelligence was the subject of Chapter 5, and the role of the expert system as a device that enables the computer to represent knowledge and perform reasoning was described. Knowledge was also a topic discussed in Chapter 6, where attention was focused on the artificial neural network and its contribution to data analysis and information processing.

The information technologies selected for inclusion in this book reflect tools and techniques that have been driven by developments outside the domain of environmental decision making. Application of these technologies to the problems and issues that fall under the umbrella of environmental decision making is relatively new and seems to fit the familiar pattern with many computer innovations where only after a technology has been proven and made widely available are its application areas fully addressed. As new applications are discovered, new technology becomes more rapidly adopted and is gradually implemented in the workplace.

Information technology is not a panacea. It is the application of scientific knowledge to satisfy a human need. Its goal is simply to help produce better decisions and to automate critical aspects of the decision-making process to enable human decision makers to concentrate on the more vexing issues surrounding a given decision problem. The information technologies presented in this book can make a valuable contribution to the decision-making process and offer the potential to automate key tasks. The chief responsibility of the decision maker in all of this is to discover the appropriate level of automation and the optimal mix of technologies that will provide the requisite support given the problems and situations that characterize the operational setting in which decisions take place.

Automated Decision Making in Perspective

Decision making involves several related yet sufficiently different tasks that range from collecting information, weighing that information, and searching for alternatives, to ultimately making a well-reasoned choice. These tasks are time-consuming, tedious, confusing, and stressful. All too often environmental decision makers operate under conditions of stress, due to crisis conditions or some type of hostile situation, or in response to other psychologically or physically constraining influences. This is reason enough to encourage the use of tools that permit the decision maker to think through a problem in a structured way that filters out extraneous noise. Yet, as discussed in Chapter 2, environmental decision making also involves a level of complexity that easily frustrates conventional problem-solving approaches. Environmental problems are multi-criteria, multivariate decision problems with complexity at all levels of the process. Complexity is introduced at one level when the decision maker must select from a number of alternatives; the more alternatives, the greater the complexity. At another level, complexity manifests itself when the decision maker must make a judgment or develop standards and criteria on which judgments are to be based and supported. Tools that help simplify complex situations, provide information and knowledge, and enhance the formulation of strategies to supplement the experience of the decision maker quickly become indispensable. Still another stress inducer is the pressure exerted on the decision-making process by time. Environmental decisions cannot wait, and they have far-reaching ecological, social, economic, and political ramifications. They must be made on a timely basis in response to the myriad influences that stand to be affected by the ultimate decision. Environmental decision makers also have an ethical obligation to be proactive. Having quick access to data—the ability to transform data into information, link information to knowledge, and execute at nearly the speed of light—plays a central role in a decision maker's ability to respond appropriately and with confidence, thereby minimizing undue delay.

Combining the factors of stress, complexity, and time provides a sound rationale for constructing automated decision aids. Automation, while not automatic, implies that some aspect of the decision problem is undertaken by machine. The machine becomes the principal device, creating an environment where the decision maker interacts with the computer to find a solution to a given problem. The level of automation defining this relationship varies depending on the degree of automation required to satisfy the motivating need.

Thus, automation is not an absolute, and the information technologies presented in this book reflect a range of automation potential. The GIS, for example, offers the capability to simplify and organize the problem of search and begins the task of converting data into information. Through the GIS, fairly well-structured problems can be explored and the spatial qualities describing a decision problem can be incorporated into the solution. The decision support system and its realization as a spatial decision support system were shown to be applicable to a class of problems that are ill-defined or ill-structured. By means of the DSS, the blending of database management with modeling creates a technology that facilitates simulation and the examination of more complex "what if" types of questions. Through the DSS, alternative solutions can be explored and the decision maker can generate and test a series of scenarios with a clearer understanding of their possible outcomes.

Expert systems were introduced as an effective way of incorporating diverse forms of knowledge, enabling the computer to automate aspects of the decision process that had previously required application of human knowledge, judgment, and experience exclusively. The expert system, when put into the decision process, is a way of supplementing the decision maker's knowledge. With an expert system, routine tasks can be performed by nonexperts, and complex tasks can be simplified. Knowledge expressed in the form of an artificial neural network is a technology capable of processing data in situations where formal rules cannot be defined. The neural network, representing a simplified construct based on the biological properties of the brain, demonstrates that a computer can be trained to recognize and process patterns in data. This unique capability provides a powerful alternative to the general problem of translating raw data into something more meaningful, particularly in situations where specialized knowledge or a detailed methodology is required but impossible to formulate.

The technologies examined in this book are integrative in nature. As suggested early in Chapter 1 and demonstrated with examples throughout the rest of the chapter, each technology can function as a stand-alone system or can be merged with a related technology to produce a hybrid system. Although the potential of hybrid systems was not examined in detail, such systems were shown to extend the capabilities of the technologies involved. Examples that point the way to a future trend in information technology include the merger of GIS with DSS to produce spatial decision support systems; the synthesis of GIS and expert systems, yielding the expert-GIS (or knowledge-based GIS); and the linking of artificial neural networks with decision support systems and GIS.

Looking Forward

As the pace of change in the technologies that handle information continues to get faster, new technologies emerge and old ones reappear in different forms. To the decision maker, these technologies must be carefully evaluated, not only in terms of their ability to significantly improve the effectiveness and quality of a decision, but also in relation to their appropriateness to the range of decision problems encountered. Because there is a natural tendency to embrace a new idea simply because it is new, failing to look critically at an innovation can contribute to unrealistic expectations and retard its adoption. Technology itself is not a solution; we need to keep a critical eye on the horizon to make certain that future technological methodologies fit the demands of environmental problem solving.

The challenge of managing information, together with the ever-present need to see relationships in data, discover new meanings, and develop a clearer understanding of process, motivate scientists and engineers to search for new methods and expand the capabilities of machines. Several emerging technologies germane to environmental decision making suggest a complement of tools and approaches that decision makers may draw upon in the very near future. Some of the more promising are:

- scientific visualization,
- qualitative reasoning,
- hyperinformation, and
- virtual reality.

Visualization

Scientific visualization was crystallized as a formal concept in an NSF commissions report produced by McCormick et al. (1987). The concept refers to the use of visual tools, usually in the form of computer graphics, to help analysts explore data and develop insight. Emphasis is typically given to three interrelated activities: (1) purposeful exploration of data, (2) searches for new patterns, and (3) development of questions and hypotheses. From the decision maker's perspective, visualization represents a shift away from numerical reasoning towards visual reasoning that draws upon simple data visualization displays in the form of charts, graphs, and maps, or more sophisticated computer graphics that apply animation and simulation techniques to create visual models of structures and processes that cannot otherwise be seen. Examples of research in visualization with important implications to environmental decision making include MacDougall (1992), Gimblett (1990), vanVoris et al. (1993), and DiBiase et al. (1992).

Qualitative Reasoning

Environmental systems, while defined quantitatively, are often qualitative and can be understood in an almost intuitive fashion. Furthermore, the behavior of a given environmental process, although often described by the exact values of a set of defining variables, cannot be adequately captured by such descriptions and still provide the level of insight necessary to gain a complete picture of system functioning. The insightful concepts and distinctions tend to be qualitative and do not lend themselves to precise measurement. In fact, given an environmental system with many sources of uncontrolled variation, it may not be possible to know the initial state of the system, the role played by external factors, or how variations are introduced through the system. Although these features of an environmental system may be sources of frustration and uncertainty, they do not preclude decision makers from reaching conclusions concerning the system's behavior. Rather, when confronted with incomplete knowledge of a system, decision makers create and use qualitative descriptions of key processes that capture important distinctions that explain unique qualitative differences and ignore others. This purely mental activity is a human reasoning process that permits inference with incomplete knowledge. Transferring this mental process to a computer introduces a new approach to modeling and simulation, called qualitative reasoning, that has tremendous potential in environmental analysis where quantitative techniques fail to offer suitable descriptions of process. The defining work in this developing approach to complex problem solving includes Kuipers (1994), Falting (1992), Bobrow (1984), and Robertson (1991). Vieu and Martin-Clouaire (1994), Lein (1993c), and Schaefer (1992) have examined qualitative reasoning from an environmental perspective and explored its potential role in environmental modeling and prediction.

Hyperinformation

The volume of data and information that can be generated and made available to a decision maker can quickly saturate the decision process and introduce unnecessary confusion. The emergence of hyperinformation is one way for all decision makers to selectively delay exposure to complexity. Although a simple definition of this broad concept has yet to crystallize, hyperinformation explains the linear and nonlinear arrangement of various forms of text, data, maps, and other media into electronic documents that can be stored and accessed in a machine environment. Such documents, referred to as hypertext, hyperdata, hypermaps, and hypermedia, are intrinsically dynamic and can include video clips, simulations, and animations, or can be calculated or generated as they are selected or requested by a user. A detailed introduction to the concept of hyperinformation as it relates to database systems can be found in Parsaye and Chignell (1993). A general treatment of the concept with reference to its application in decision support and GIS can be found in Cassettari (1993).

Virtual Reality

Although presently a technology still very much on the research and development frontier, virtual reality offers a unique approach to decision making that facilitates the representation of a problem in the form of a metal context-model that may be experienced firsthand within a computer environment. Through the development of a virtual reality interface, a decision maker's time and space can be matched to the time and space dimension describing the problem. This "matching" of two contrasting realities allows the decision maker to "physically" visit the problem site and experience its dominating features.

Virtual reality systems also offer the potential to create a virtual world characterizing a given environmental problem that can be experienced for a wider range of purposes. In either instance, a virtual representation of a given decision problem is produced that facilitates a very different type of data visualization. Virtual representations also provide a level of synthesis that combines the problem, its features, and the decision maker in a way that present information technologies cannot. To the environmental decision maker, virtual reality may be one means of explaining and experiencing complex environmental issues. Through such a technology, problems such as the evaluation of cumulative environmental impact, risk, and developmental suitability may be examined in a manner that better illuminates the appropriate solution. Excellent references on this topic and its potential applications in decision making include Kalasky (1993), Cotton (1993), Larijani (1993), and Earnshaw, Gigante, and Jones (1993).

Final Thoughts

Information technology and environmental problems are constantly changing. Managing the entropic influences of human societies and reconciling the political, economic, social, and environmental issues that direct and motivate change on Earth place renewed emphasis on decision making and force a more critical examination of its role in the activities that manage and distribute vital environmental resources. Planning and decision making, however, cannot take place without information as a guide. Yet the information so critical to effective planning and decision making is diverse and demands careful management.

Managing environmental information introduces a range of technologies that facilitate access, storage, and analysis of data used in making decisions. Information is the key ingredient that supports efforts aimed at directing and accommodating change and preserving environmental quality. With a global population of nearly 5.6 billion people, a population growth rate estimated at 1.7 percent, and a universal desire for continuously improving living standards, tomorrow's world will be a very challenging place. Meeting tomorrow's challenges will require wisdom and creativity as well as technology capable of supporting that wisdom and creativity in new and unique ways. The information technologies discussed in this book are a cross section of methods with demonstrated utility for addressing environmental problems. As with any tool or approach, the user must learn how to employ them correctly and understand their limitations.

The aim of this book was to present a set of essential information technologies for the environmental professional, not as abstractions or exotic techniques that stand in isolation, but as complementary systems that can be linked together to help in making environmental decisions. Technology is not the end. It is a means of connecting fragments of data into a vision of tomorrow's world on which decisions are based. Our vision must, therefore, be clear because those decisions will affect the world our children inherit.

References

Aageenbrug, R. 1991. A Critique of GIS. In *Geographical Information Systems: Principles and Applications*, ed. D. Maguire, M. Goodchild, and D. Rhind. London: Longman Group, Ltd.

Altman, D. 1994. Fuzzy Set Theoretic Approaches for Handling Imprecision in Spatial Analysis. *International Journal of Geographic Information Systems* 8:271–89.

Anadalingam, G., and Westfall, M. 1988. Selection of Hazardous Waste Disposal Alternative Using Multi-Attribute Utility Theory and Fuzzy Set Analysis. *Journal of Environmental Systems* 18:69–85.

Anderson, J. 1985. *Cognitive Psychology*. New York: W.H. Freeman and Co.

Anselin, L., and Getis, A. 1992. Spatial Statistical Analysis and Geographic Information Systems. *Annals of Regional Science* 26:19–33.

Antenucci, J. et al. 1991. *Geographic Information Systems: A Guide to the Technology*. New York: Van Nostrand Reinhold.

Apostolakis, G. 1990. The Concept of Probability in Safety Assessments of Technological Systems. *Science* 250:1359–64.

Armstrong, M., Rushton, G., and Honey, R. 1991. Decision Support for Regionalization. *Computers, Environment, and Urban Systems* 15:37–53.

Armstrong, M., De, S., and Densham, P. 1990. A Knowledge-based Approach for Supporting Locational Decision Making. *Environment and Planning B: Planning and Design* 17:341–64.

Aronoff, S. 1991. *Geographic Information Systems: A Management Perspective*. Ottawa: WDL Publications.

Bailey, D., and Thompson, D. 1990. Developing Neural-Network Applications. *AI Expert* (September), 34–41.

Balachandran, C., and Fisher, P. 1990. DESERT: A Prototype Expert System to Advise on Land Degradation Control. *Land Degradation and Rehabilitation* 2:27–41.

Banai, R. 1993. Fuzziness in Geographical Information Systems. *International Journal of Geographical Information Systems* 7:315–29.

Bartell, S., Gardnet, R., and O'Neill, R. 1992. *Ecological Risk Estimation*. Boca Raton, Fla.: Lewis Publishers.

Bellman, R., and Zadeh, L. 1970. Decision Making in a Fuzzy Environment. *Management Science* 17:141–64.

Benediktsson, J., Swain, P., and Ersoy, O. 1990. Neural Network Approaches Versus Statistical Methods in Classification of Multisource Remote Sensing Data. *IEEE Transactions on Geoscience and Remote Sensing* 28:540–52.

Bennett, R., and Chorley, R. 1978. *Environmental Systems: Philosophy, Analysis and Control.* London: Methuen.

Bennett, R., and Estall, R. 1991. *Global Change and Challenge*. London: Routledge.

Beroggi, G., and Wallace, W. 1992. Real-time Control of the Transportation of Hazardous Materials. *URISA Journal* 4:56–64.

Berry, J. 1995. *Spatial Reasoning for Effective GIS*. Boulder, Colo.: GIS World.

Berry, J. 1987. Fundamental Operations in Computer Assisted Map Analysis. *International Journal of Geographical Information Systems* I:119–36.

Bezdek, J., ed. 1987. *Analysis of Fuzzy Information*. Boca Raton, Fla.: CRC Press.

Bidogli, H. 1984. *Decision Support Systems: Principles and Practice*. St. Paul, Minn.: West Publishing, Co.

Bobrow, D. 1984. Qualitative Reasoning About Physical Systems. Cambridge, Mass.: MIT Press.

Bonano, E. et al. 1989. *The Use of Expert Judgments in Performance Assessment of HLW Repositories* (SAND 89-0495C). Albuquerque, N. Mex.: Sandia National Laboratories.

Bonczek, R., Holsapple, C., and Whinston, A. 1984. Developments in Decision Support Systems. *Advances in Computers* 23:1141–75.

Bonham-Carter, G. 1994. *Geographic Information Systems for Geoscientists: Modeling with GIS*. Tarrytown, N.Y.: Pergamon/Elsevier Science Publications.

Boznar, M. 1993. A Neural Network-based Method for Short-Term Predictions of Ambient SO^2 Concentrations. *Atmospheric Environment* 27B:221–31.

Brouwer, F. 1987. *Integrated Environmental Modeling: Design and Tools*. Dordrecht, Netherlands: Kluwer Academic Publishers.

Brown, S. et al. 1994. Linking Multiple Accounts with a GIS as Decision Support Systems to Resolve Forestry/Wildlife Conflicts. *Journal of Environmental Management* 42:349–64.

Buchanan, B., and Duda, R. 1983. Principles of Rule-based Expert Systems. *Advances in Computers* 22:163–216.

Budic, Z., and Giodschalk, D. 1994. Implementation and Management Effectiveness in Adoption of GIS Technology in Local Governments. *Computers, Environment, and Urban Systems* 18:285–304.

Burrough, P. 1988. *Principles of Geographical Information Systems for Land Resource Assessment*. Oxford: Clarendon Press.

Caldwell, L. 1987. The Contextual Basis for Environmental Decision Making. *The Environmental Professional* 9:302–8.

Campbell, H. 1994. How Effective Are GIS in Practice? *International Journal of Geographical Information Systems* 8:309–25.

Campbell, J. 1987, *Introduction to Remote Sensing*. New York: Guilford Press.

Carrara, A., and Guzzetti, F. 1995. *Geographic Information Systems in Assessing Natural Hazards*. Dordrecht, Netherlands: Kluwer Academic Publishers.

Carroll, J., and Johnson, E. 1990. *Decision Research: A Field Guide*. Applied Social Research Methods Series, vol. 22. Newbury Park, Calif.: Sage Publications.

Cartwright, T. 1993. *Modeling the World in a Spreadsheet*. Baltimore, Md.: Johns Hopkins University Press.

Carver, S. 1991. Integrating Multi-Criteria Evaluation with Geographical Information Systems. *International Journal of Geographical Information Systems* 5:321–39.

Cassettari, S. 1993. *Introduction to Integrated Geo-Information Management*. London: Chapman and Hall.

Castillo, E., and Alvarez, E. 1990. Uncertainty Methods in Expert Systems. *Microcomputers in Civil Engineering* 5:43–58.

Chang, L., and Burrough, P. 1987. Fuzzy Reasoning: A New Quantitative Aid for Land Evaluation. *Soil Survey and Land Evaluation* 7:69–80.

Chechile, R. 1991. Introduction to Environmental Decision Making. In *Environmental Decision Making: A Multidisciplinary Perspective*, ed. R. Chechile and S. Carlisle. New York: Van Nostrand Reinhold.

Chen, J., et al. 1994. The Development of a Knowledge-based Geographic Information System for the Zoning of Rural Areas. *Environment and Planning B: Planning and Design* 21:179–90.

Chester, M. 1993. *Neural Networks: A Tutorial.* Englewood Cliffs, N.J.: PTR Prentice-Hall.

Chorley, R., and Kennedy, B. 1971. *Physical Geography: A Systems Approach.* London: Prentice-Hall International.

Civco, D. 1993. Artificial Neural Networks for Land Cover Classification and Mapping. *International Journal of Geographical Information Systems* 7:173–86.

Clark, W. 1989. The Human Ecology of Global Change. *International Social Science Journal* 41:315–45.

Clarke, M. 1989. Geography and Information Technology: Blueprint for a Revolution? In *Horizons in Human Geography,* ed. D. Gregory. New York: Barnes and Noble.

Cocklin, C., Parker, S., and Hay, J. 1992. Notes on Cumulative Environmental Change I: Concepts and Issues. *Journal of Environmental Management* 35:31–49.

Cohen, A. 1984. Overview and Definition of Risk. *Environment International* 10:359–66.

Connor, S., and Allen, P. 1994. Policy and Decision Support in Sustainable Development Planning. *Project Appraisal* 9:95–103.

Constantine, M., and Ulvila, J. 1990. Testing Knowledge-based Systems. *Expert Systems with Applications* 1:237–48.

Cook, D., and Wolf, M. 1991. A Back-Propagation Neural Network to Predict Average Air Temperature. *AI Applications in Natural Resource Management* 5:40–46.

Cotton, B. 1993. *Understanding Hypermedia.* London: Phaidon Press.

Coughlin, J., and Running, S. 1989. An Expert System to Aggregate Biophysical Attributes of a Forested Landscape Within a GIS. *AI Applications in Natural Resource Management* 3:35–43.

Covello, V. et al. 1987. *Uncertainty in Risk Assessment, Risk Management, and Decision Making.* New York: Plenum Press.

Crosslin, R. 1991. Decision-Support Methodology for Planning and Evaluating Public-Private Partnerships. *Journal of Urban Planning and Development* 117:15–31.

Cruz, C. 1989. Artificial Neural Networks: A High Level Perspective. In *Neural Networks at a Glance,* ed. R. Morley. Amherst, N.H.: Publishing Corp.

Davies, C., and Medyckyj-Scott, D. 1994. GIS Usability: Recommendations Based on the User's View. *International Journal of Geographical Information Systems* 8:175–89.

Davies, R., and Lein, J. 1991. Applying an Expert System Methodology for Solid Waste Landfill Site Selection. *URISA Proceedings,* vol. 1, 40–53.

Davis, J., and Clark, J. 1989. A Selective Bibliography of Expert Systems in Natural Resource Management. *AI Applications in Natural Resource Management* 3:1–18.

Davis, J., Compagnoni, P., and Nanninga, P. 1987. Roles for Knowledge-based Systems in Environmental Planning. *Environment and Planning B: Planning and Design* 14:239–54.

Davis, J., and Grant, I. 1987. ADAPT: A Knowledge-based Decision Support System for Producing Zoning Schemes. *Environment and Planning B: Planning and Design* 14:53–66.

Davis, M. 1988. *Applied Decision Support.* Englewood Cliffs, N.J.: Prentice-Hall.

Dayhof, J. 1990. *Neural Network Architectures.* New York: Van Nostrand Reinhold.

Debons, A., Horne, E., and Cronenweth, S. 1988. *Information Science: An Integrated View.* Boston, Mass.: G.K. Hall and Co.

Densham, P. 1991. Spatial Decision Support Systems. In *Geographic Information Systems: Principles and Applications,* ed. D. Maguire, M. Goodchild, and D. Rhind. New York: Wiley and Sons.

Densham, P., and Goodchild, M. 1989. Spatial Decision Support Systems. Proceedings of GIS/LIS '89, ACSM, Bethesda, Md., pp. 707–16.

Derr, V., and Slutz, R. 1994. Prediction of El Nino Events in the Pacific by Means of Neural Networks. *AI Applications in Natural Resource Management* 8:51–63.

DiBiase, D. et al. 1992. Animation and the Role of Map Design in Scientific Visualization. *Cartography and Geographic Information Systems* 19:201–14.

Dowd, P. 1994. A Neural Network Approach to Geostatistical Simulation. *Mathematical Geology* 26:491–503.

Dreyer, P. 1993. Classification of Land Cover Using Optimized Neural Networks of SPOT Data. *Photogrammetric Engineering and Remote Sensing* 59:617–21.

Dunn, R., Harrison, A., and White, J. 1990. Positional Accuracy and Measurement Error in Digital Databases of Land Use: An Empirical Study. *International Journal of Geographical Information Systems* 4:385–98.

Dyer, R. 1989. Adapting Expert Systems to Multiple Locations. *AI Applications in Natural Resource Management* 3:11–16.

Earnshaw, R., Gigante, M., and Jones, H. 1993. *Virtual Reality Systems*. London: Academic Press.

Eastman, R., and Fulk, M. 1993. Long Sequence Time Series Evaluations Using Standardized Principle Components. *Photogrammetric Engineering and Remote Sensing* 59:1307, 1313–14.

Eastman, R. et al. 1993. GIS and Decision Making. *Explorations in Geographic Information System Technology* 4. Geneva: UNITAR.

Eastman, R. et al. 1995. Raster Procedures for Multi-Criteria/Multi-Objective Decisions. *Photogrammetric Engineering and Remote Sensing* 61:539–47.

Edamura, T., and Kawai, T. 1991. An Expert System for Forecasting Roadside Development. *Computers, Environment, and Urban Systems* 15:141–50.

Falting, B. 1992. *Recent Advances in Qualitative Physics*. Cambridge, Mass.: MIT Press.

Farmer, D., and Rycroft, M. 1991. *Computer Modelling in the Environmental Sciences*. Oxford: Clarendon Press.

Fedra, K., and Reitsma, R. 1990. Decision Support and Geographical Information Systems. In *Geographical Information Systems for Urban and Regional Planning*, ed. H. Scholten and J. Stillwell. Dordrecht, Netherlands: Kluwer Academic Publishers.

Festinger, L. 1964. *Conflict, Decision and Dissonance*. Stanford, Calif.: Stanford University Press.

Fetzer, J. 1990. *Artificial Intelligence: Its Scope and Limits*. Dordrecht, Netherlands: Kluwer Academic Press.

Fields, D., and Kim, T. 1992. Application of a Computer-aided Expert Decision Support System to Rural Development in Kenya. *Computers, Environment, and Urban Systems* 16:415–33.

Fiksel, J. 1990. Risk Models for Wildfowl Diseases. *Journal of Environmental Management* 30:371–80.

Finn, J. 1993. Use of the Average Mutual Information Index in Evaluating Classification Error and Consistency. *International Journal of Geographical Information Systems* 7:349–66.

Fischer, M. 1994. From Conventional to Knowledge-based Geographic Information Systems. *Computers, Environment, and Urban Systems* 18:233–42.

Fishman, G., and Kivat, P. 1967. The Analysis of Simulation Generated Time Series. *Management Sciences* 13:121–39.

Fisher, P. 1991. Modelling Soil Map-Unit Inclusion by Monte Carlo Simulation. *International Journal of Geographical Information Systems* 5:193–208.

Fishwick, P. 1995. Simulation Model Design and Execution. Englewood Cliffs, N.J.: Prentice-Hall.

Fleming, P. 1991. Expert Judgment and High-Level Nuclear Waste Management. *Policy Studies Review* 11:311–31.

Foody, G. 1995. Land Cover Classification by an Artificial Neural Network with Ancillary Information. *International Journal of Geographical Information Systems* 9:527–42.

French, M. 1992. Rainfall Forecasting in Space and Time Using a Neural Network. *Journal of Hydrology* 137:1–31.

Frenkiel, F., and Goodall, D. 1978. *Simulation Modelling of Environmental Problems, SCOPE 9*. Chichester, U.K.: Wiley and Sons.

Gaines, B. 1984. Fundamentals of Decison: Probabilistic, Possibilistic, and Other Forms of Uncertainty in Decision Analysis. In *Fuzzy Sets and Decision Analysis*, ed. H. Zimmerman. Amsterdam: North-Holland.

Gallopin, G. 1991. Human Dimensions of Global Change: Linking the Global and the Local Processes. *International Social Sciences Journal* 43:707–18.

Gardels, K. 1987. The Expert Geographic Knowledge System. *Proceedings, Auto-Carto* 8:520–9.
Geraghty, P. 1992. Environmental Assessment and the Application of an Expert Systems Approach. *Town Planning Review* 63:123–42.
Giarratano, J., and Riley, G. 1989. *Expert Systems: Principles and Programming*. Boston: PWS-Kent Publishing Co.
vanGigch, J. 1991. *System Design Modeling and Metamodeling*. New York: Plenum Press.
Gimblett, R. 1990. Visualizations: Linking Dynamic Spatial Models with Computer Graphic Algorithms. *URISA Journal* 2:26–34.
Gimblett, R., and Ball, G. 1995. Neural Network Architectures for Monitoring and Simulating Changes in Forest Resource Management. *AI Applications in Natural Resource Management* 9:103–18.
Gong, P. 1996. Integrated Analysis of Spatial Data from Multiple Sources: Using Evidential Reasoning and Artificial Neural Network Techniques for Geologic Mapping. *Photogrammetric Engineering and Remote Sensing* 62:513–23.
Goodchild, M. 1992. Geographical Data Modeling. *Computers and Geosciences* 18:401–8.
Goodchild, M., and Gopal, S., eds. 1989. *Accuracy of Spatial Databases*. London: Taylor and Francis.
Goodchild, M., Parks, B., and Steyaert, L. 1993. *Environmental Modeling with Geographic Information Systems*. Oxford: Oxford University Press.
Gordon, S. 1985. *Computer Models in Environmental Planning*. New York: Van Nostrand Reinhold.
Gould, M. 1989. The Value of Spatial Decision Support Systems for Oil and Chemical Spill Response. Proceedings of the 12th Applied Geography Conference, Binghamton, N.Y., pp. 75–83.
Gorry, G., and Scott-Morton, M. 1971. Framework for Management Information Systems. *Sloan Management Review* 13:55–70.
Graybeal, W., and Pooch, U. 1980. *Simulation: Principles and Methods*. Boston: Little, Brown and Co.
Greenwell, M. 1988. *Knowledge Engineering for Expert Systems*. New York: Wiley and Sons.
Grossman, W., and Eberhardt, S. 1992. Geographic Information Systems and Dynamic Modeling. *Annals of Regional Science* 26:53–66.
Gupta, M., and Qi, J. 1992. On Fuzzy Neuron Models. In *Fuzzy Logic for the Management of Uncertainty*, ed. L. Zadeh and J. Kacprzyk. New York: Wiley and Sons.

Haines-Young, R., Green, D., and Cousins, S. 1994. Landscape Ecology and GIS. London: Taylor and Francis.
Han, S., and Kim, T. 1989. Can Expert Systems Help with Planning? *American Planning Association Journal* 34:296–308.
Han, S., Kim, T., and Adiguzel, I. 1991. XPlanner: A Knowledge-based Decision Support System for Facility Management and Planning. *Environment and Planning B: Planning and Design* 18:205–24.
Hannon, B., and Ruth, M. 1994. *Dynamic Modeling*. New York: Springer-Verlag.
Hardisty, J., Taylor, D., and Metcalfe, S. 1993. *Computerized Environmental Modeling*. Chichester, U.K.: Wiley and Sons.
Hazelton, N., Leahy, F., and Williamson, I. 1992. Integrating Dynamic Modeling and GIS. *URISA Journal* 4:47–58.
Heermann, P., and Khanzeni, N. 1992. Classification of Multispectral Remote Sensing Data Using a Back-Propagation Neural Network. *IEEE Transactions on Geoscience and Remote Sensing* 30:81–88.
Henderson, J. 1987. Finding Synergy Between Decision Support Systems and Expert Systems Research. *Decision Sciences* 18:333–49.
Hepner, G. et al. 1990. Artificial Neural Network Classification Using a Minimal Training Set. *Photogrammetric Engineering and Remote Sensing* 56:469–73.
Hertz, J., Krogh, A., and Palmer, R. 1991. *Introduction to the Theory of Neural Computation*. Redwood City, Calif.: Addison-Wesley.
vanHerwijen, M., Janssen, R., and Nijkamp, P. 1993. A Multi-Criteria Decision Support Model and GIS for Sustainable Development Planning. *Project Appraisal* 8:9–23.

Heuvelink, G., and Burrough, P. 1989. Propagation of Errors in Spatial Modelling with GIS. *International Journal of Geographical Information Systems* 3:303–22.

Heuvelink, G., and Burrough, P. 1993. Error Propagation in Cartographic Modelling Using Boolean Logic and Continuous Classification. *International Journal of Geographic Information Systems* 7:231–46.

Heydinger, R., and Zenter, R. 1983. Multiple Scenario Analysis: Introducing Uncertainty into the Planning Process. In *Applying Methods and Techniques of Futures Research,* ed. J. Morrison. San Francisco: Jossey-Bass, Inc.

Hinton, G. 1989. Connectionist Learning Procedures. *Artificial Intelligence* 40:185–234.

Holmberg, S. 1994. Geoinformatics for Urban and Regional Planning. *Environment and Planning B: Planning and Design* 21:5–19.

Holsapple, C. 1983. The Knowledge System for a Generalized Problem Processor. Krannert Institute Paper, Purdue University, West Lafayette, Ind.

Holsapple, C., and Whinston, A. 1983. Software Tools for Knowledge Fusion. *Computerworld* 17:38–47.

Hopple, G. 1988. *The State of the Art in Decision Support Systems* Wellesley, Mass.: QED Information Sciences, Inc.

Horikawa, S., Furuhashi, T., and Uchikawa, Y. 1992. On Fuzzy Modeling Using Fuzzy Neural Networks. *IEEE Transactions on Neural Networks* 3:801–6.

Horsak, R., and Damica, S. 1985. Selection and Evaluation of Hazardous Waste Disposal Sites Using Fuzzy Set Analysis. *Journal of the Air Pollution Control Association* 35:1081–5.

Huggett, R. 1980. *Systems Analysis in Geography*. London: Clarendon Press.

Huggett, R. 1993. Modelling the Human Impact on Nature. Oxford: Oxford University Press.

Hunsaker, C., Graham, R., and Suter, G. 1990. Assessing Ecological Risk on a Regional Scale. *Environmental Management* 14:325–32.

Hushon, J. 1990. *Expert Systems for Environmental Applications*. ACS Symposium Series 431. Washington, D.C.: American Chemical Society.

Hutchinson, C. 1992. Natural Resource and Environmental Information for Decisionmaking. Washington, D.C.: World Bank Publications.

Huxhold, W. 1991. *An Introduction to Urban Geographic Information Systems*. Oxford: Oxford University Press.

Huxhold, W., and Levinsohn, A. 1995. *Managing Geographic Information Systems Projects*. Oxford: Oxford University Press.

Inbar, M. 1979. *Routine Decison-Making Process*. Beverly Hills, Calif.: Sage Publications.

Jackson, P. 1990. *Introduction to Expert Systems*. Wokingham, U.K.: Addison-Wesley.

Jakeman, A., Beck, M., McAleer, M. 1993. *Modelling Change in Environmental Systems*. New York: Wiley and Sons.

Jankowski, P. 1995. Integrating Geographical Information Systems and Muliple-Criteria Decision-Making Methods. *International Journal of Geographical Information Systems* 9:251–73.

Jeffers, J. 1982. *Modelling, Outline Studies in Ecology*. London: Chapman and Hall.

Jensen, J. 1996. *Introductory Digital Image Processing*. Englewood Cliffs, N.J.: Prentice-Hall.

Johnson, P. 1983. What Kind of an Expert Should a System Be? *Journal of Medicine and Philosophy* 8:77–97.

Jorgensen, S., Nielsen, S., and Sorensen, B. 1995. *Handbook of Environmental and Ecological Modelling*. Boca Raton, Fla.: Lewis Publishers.

Julien, B., Fenves, S., and Small, M. 1992. Knowledge Acquisition Methods for Environmental Evaluation. *AI Applications in Natural Resource Management* 6:1–20.

Kacprzyk, J. 1982. Multistage Decision Processes in a Fuzzy Environment. In *Fuzzy Information and Decision Processes,* ed. M. Gupta. Amsterdam: North-Holland.

Kalasky, R. 1993. *The Science of Virtual Reality and Virtual Environments*. Wokingham, U.K.: Addison-Wesley.

Kangari, R., and Boyer, L. 1992. Knowledge-based Systems and Fuzzy Sets in Risk Management. *Microcomputers in Civil Engineering* 2:273–83.
Kaplan, S. 1984. The Industrialization of Artificial Intelligence. *AI Magazine* (Summer), 16–21.
Keller, J., Krishnapuram, R., and Rhee, F. 1992. Evidence Aggregation Networks for Fuzzy Logic Inference. *IEEE Transactions on Neural Networks* 3:761–9.
Kim, T., Wiggins, L., and Wright, J. 1990. *Expert Systems: Applications to Urban Planning*. New York: Springer-Verlag.
Klein, M., and Methlie, L. 1990. *Expert Systems: A Decision Support Approach*. Wokingham, U.K.: Addison-Wesley.
King, D. 1990. Intelligent Decision Support. *Expert Systems with Applications* 1:23–38.
Klir, G., and Folger, T. 1988. *Fuzzy Sets, Uncertainty, and Information*. Englewood Cliffs, N.J.: Prentice-Hall.
Kliskey, A. 1995. The Role and Functionality of GIS as a Planning Tool in Natural Resource Management. *Computers, Environment, and Urban Systems* 19:15–22.
Klosterman, R., Brail, R., and Bossard, E. 1993. *Spreadsheet Models for Urban and Regional Analysis*. New Brunswick, N.J.: Center for Urban Policy Research.
Kofron, C. 1989. AI and GIS Connections. Paper Presented at GIS/LIS '89, Orlando, Fla.
Kohonen, T. 1988. An Introduction to Neural Computing. *Neural Networks* 1:3–16.
Kolias, V., and Voliotis, A. 1991. Fuzzy Reasoning in the Development of Geographical Information Systems. *International Journal of Geographical Information Systems* 5:209–23.
Konsynski, B. 1992. Review and Critique of DSS. In *Information Systems and Decision Processes*, ed. E. Stohr. Los Alamitos, Calif.: IEEE Computer Society Press.
Koplik, C., Kaplan, M., and Ross, B. 1982. The Safety of Repositories for Highly Radioactive Wastes. *Reviews of Modern Physics* 54:93–117.
Kuipers, B. 1994. *Qualitative Reasoning*. Cambridge, Mass.: MIT Press.

Labadie, J. 1986. Computerized Decision Support Systems for Water Managers. *Journal of Water Resource Planning and Management* 112:299–306.
Lagendorf, R. 1985. Computers and Decision Making. *Journal of the American Planning Association* 27:422–33.
Lanter, D., and Veregin, H. 1992. A Research Paradigm for Propagating Error in Layer-based Geographic Information Systems. *Photogrammetric Engineering and Remote Sensing* 58: 526–33.
Larijani, C. 1993. *The Virtual Reality Primer.* New York: McGraw-Hill.
Laurini, R., and Thompson, D. 1992. *Fundamentals of Spatial Information Systems*. New York: Academic Press.
Lehr, J. 1990. *Toxicological Risk Assessment*. Dublin, Ohio: The American Ground Water Trust.
Lein, J. 1989. An Expert System Approach to Environmental Impact Assessment. *International Journal of Environmental Studies* 33:13–27.
———. 1990. Exploring a Knowledge-based Procedure for Developmental Suitability Analysis. *Applied Geography* 10:171–86.
———. 1992a. Expressing Environmental Risk Using Fuzzy Variables. *The Environmental Professional* 14:257–67.
———. 1992b. Modeling Environmental Impact Using an Expert-Geographic Information System. Proceedings: GIS/LIS '92, San Jose, Calif., pp. 436–44.
———. 1993a. Applying Expert System Technology to Carrying Capacity Assessment: A Demonstration Prototype. *Journal of Environmental Management* 37:63–84.
———. 1993b. Formalizing Expert Judgment in the Environmental Impact Assessment Process. *The Environmental Professional* 15:95–102.
———. 1993c. Qualitative Reasoning and Environmental Decision Making. In *Papers and Proceedings of the Applied Geography Conferences*, ed. J. Frazier, 16:130–35.
———. 1995. Mapping Environmental Carrying Capacity Using an Artificial Neural Network. *Land Degradation and Rehabilitation* 6:17–28.
———. (in press). Linkage and Scale Effects in the Study of Global Environmental Change. *The Environmental Professional.*

Leung, Y. 1988. *Spatial Analysis and Planning Under Imprecision.* Amsterdam: North-Holland.

———. 1994. Inference with Spatial Knowledge: An Artificial Neural Network Approach. *Geographical Systems* 1:103–21.

Lewandowski, A. 1991. *Methodology, Implementation, and Application of Decision Support Systems.* New York: Springer-Verlag.

Lindley, D. 1982. Scoring Rules and the Inevitability of Probability. *International Statistical Review* 50:1–26.

Lippman, R. 1987. An Introduction to Computing with Neural Nets. *IEEE Acoustics, Speech, and Signal Processing Magazine* (April), 4–22.

Lodwick, W., Monson, W., Svoboda, L. 1990. Attribute Error and Sensitivity Analysis of Map Operations in Geographical Information Systems. *International Journal of Geographical Information Systems* 4:413–28.

Loh, D., and Rykiel, E. 1992. Integrated Resource Management Systems. *Environmental Management* 16:167–77.

Lunardhi, A., and Passino, K. 1995. Verification of Qualitative Properties of Rule-based Expert Systems. *Applied Artificial Intelligence* 9:587–621.

MacDougall, E. 1992. Exploratory Analysis, Dynamic Statistical Visualization, and Geographic Information Systems. *Cartography and Geographic Information Systems* 19:237–46.

Maguire, D. 1991. An Overview and Definition of GIS. In *Geographical Information Systems: Principles and Applications,* ed. D. Maguire, M. Goodchild, and D. Rhind. New York: Wiley and Sons.

Maidment, D., and Djokic, D. 1990. Creating an Expert Geographic Information System. (Unpublished manuscript.)

Mann, C., and Hunter, R. 1988. Probabilities of Geologic Events and Processes in Natural Hazards, Zeitschrt. *Geomorphology, N.F.* 67:39–52.

Mantara, R. 1990. *Approximate Reasoning Models.* Chichester, U.K.: Ellis Horwood, Ltd.

Masser, I., and Blackmore, M. 1991. *Handling Geographic Information.* London: Longman.

Matheus, C., and Hohensee, W. 1987. Learning in Artificial Neural Systems. *Computer Intelligence* 3:283–94.

McClean, C., Watson, P., and Wadsworth, A. 1995. Land Use Planning: A Decision Support System. *Journal of Environmental Planning and Management* 38:77–90.

McCloy, K. 1995. *Resource Management Information Systems: Process and Practice.* New York: Taylor and Francis.

McCormick, B., DeFanti, T., and Brown, M. 1987. Visualization in Scientific Computing. Report to the National Science Foundation, Panel on Graphics, Image Processing, and Workstations, Baltimore, Md.

McGinnis, D. 1994. Predicting Snowfall from Synoptic Circulation. In *Neural Nets: Applications in Geography,* ed. B. Hewitson and R. Crane. Dordrecht, Netherlands: Kluwer Academic Press.

McGraw, K., and Harbison-Briggs, K. 1989. *Knowledge Acquisition: Principles and Guidelines.* Englewood Cliffs, N.J.: Prentice-Hall.

McLaughlin, J., and Coleman, D. 1990. Land Information Management into the 1990's. *World Cartography* 20:103–25.

Medsker, L., and Liebowitz, J. 1994. *Design and Development of Expert Systems and Neural Networks.* New York: Macmillan Publishing Co.

Mehra, P., and Wah, B. 1992. *Artifical Neural Networks: Concepts and Theory.* IEEE Los Alamitos, Calif.: Computer Society Press.

Mercer, K. 1995. An Expert System Utility for Environmental Impact Assessment in Engineering. *Journal of Environmental Management* 45:1–23.

Miller, A. 1993. The Role of Analytical Science in Natural Resource Decision Making. *Environmental Management* 17:563–74.

Mitta, S. 1986. *Decision Support Systems: Tools and Techniques.* New York: Wiley and Sons.

Moffatt, I. 1990. The Potentialities and Problems Associated with Applying Information Technology to Environmental Management. *Journal of Environmental Management* 30:209–20.

Morgan, M., and Henruion, M. 1990. *Uncertainty: A Guide to Dealing with Uncertainty in Quantitative Risk and Policy Analysis.* Cambridge, U.K.: Cambridge University Press.

Morgan, M. et al. 1985. Uncertainty in Risk Assessment. *Environment, Science, and Technology* 19:662–67.

Morse, B. 1987. Expert System Interface to a GIS. Proceeding, *Auto-Carto* 8:535–41.

Mulder, J., and Corns, I. 1995. NAIA: A Decision Support System for Predicting Ecosystems from Existing Land Resource Data. *AI Applications in Natural Resource Management* 9:49–61.

Neelamkavil, F. 1986. *Computer Simulation and Modeling.* New York: Wiley and Sons.

Nelson, M., and Illingworth, W. 1991. *A Practical Guide to Neural Nets.* Reading, Mass.: Addison-Wesley.

Newcomer, J., and Szajgin, J. 1984. Accumulation of Thematic Map Errors in Digital Overlay Analysis. *The American Cartographer* 11:58–62.

O'Keefe, R., Balci, O., and Smith, E. 1988. Validating Expert System Performance. *AI Applications in Natural Resource Management* 2:35–43.

O'Neill, R. 1988. Hierarchy Theory and Global Change. In *Scales and Global Change,* ed. T. Rosswell. New York: Wiley and Sons.

O'Neill, R. et al. 1986. *Hierarchical Concept of Ecosystems.* Princeton, N.J.: Princeton University Press.

Openshaw, S. 1990. Spatial Analysis and Geographic Information Systems: A Review of Progress and Possibilities. In *Geographical Information Systems for Urban and Regional Planning,* ed. H. Scholten and J. Stillwell. Dordrecht, Netherlands: Kluwer Academic Publishers.

Openshaw, S. 1994. Neuroclassification of Spatial Data. In *Neural Nets: Applications in Geography,* ed. B. Hewitson and R. Crane. Dordrecht, Netherlands: Kluwer Academic Publishers.

Openshaw, S., Charlton, M., and Carver, S. 1991. Error Propagation: A Monte Carlo Simulation. In *Handling Geographical Information,* ed. I. Masser and M. Blackmore. London: Longman.

Pal, S., and Mitra, S. 1992. Multilayer Perception, Fuzzy Sets, and Classification. *IEEE Transactions on Neural Networks* 3:683–97.

Parsaye, K., and Chignell, M. 1988. *Expert Systems for Experts.* New York: Wiley and Sons.

Parsaye, K., and Chignell, M. 1993. *Intelligent Database Tools and Applications.* New York: Wiley and Sons.

Pattie, D., and Haas, G. 1996. Forecasting Wilderness Recreation Use: Neural Networks vs. Regression. *AI Applications in Natural Resource Management* 10:67–76.

Payne, E., and McArthur, R. 1990. *Developing Expert Systems.* New York: Wiley and Sons.

Petak, W., and Atkisson, A. 1982. *Natural Hazard Risk Assessment and Public Policy.* New York: Springer-Verlag.

Peuquet, D. 1984. A Conceptual Framework and Comparison of Spatial Data Models. *Cartographica* 21:66–113.

Pfeifer, R., and Luthi, H. 1987. Decision Support Systems and Expert Systems. In *Expert Systems and Artificial Intelligence in Decision Support,* ed. H. Sol. Boston: D. Reidel Publishing.

Plant, R. 1993. Expert Systems in Agriculture and Resource Management. *Technological Forecasting and Social Change* 43:241–57.

Rich, E. 1983. *Artificial Intelligence.* New York: McGraw-Hill.

Roberts, N. 1994. The Global Environmental Future. In *The Changing Global Environment,* ed. N. Roberts. Oxford: Blackwell Science.

Robertson, D. 1991. *Eco-logic: Logic-Based Approaches to Ecological Modeling.* Cambridge, Mass.: MIT Press.

Robinson, J. 1965. A Machine-Oriented Logic Based on the Resolution Principle. *Journal of the Association for Computing Machinery* 12:23–41.

———. 1991. Modeling the Interactions Between Human and Natural Systems. *International Social Sciences Journal* 43:629–47.

Robinson, V. 1988. Some Implications of Fuzzy Set Theory Applied to Geographic Databases. *Computers, Environment, and Urban Systems* 12:89–97.

Rogerson, P., and Fotheringham, S. 1994. GIS and Spatial Analysis: Introduction and Overview. In *Spatial Analysis and GIS,* ed. S. Fotheringham and S. Rogerson. London: Taylor and Francis.

Ross, B. 1989. Scenarios for Repository Safety Analysis. *Engineering Geology* 26:285–99.

Rowe, W. 1977. *An Anatomy of Risk.* New York: Wiley and Sons.

Rumelhart, D. 1986. *Parallel Distributed Processing, Volume 1.* Cambridge, Mass.: MIT Press.

Saaty, T. 1978. Exploring the Interface Between Hierarchies, Multiple Objectives, and Fuzzy Sets. *Fuzzy Sets and Systems* 1:57–64.

Sage, A. 1991. Decision Support Systems Engineering. New York: Wiley and Sons.

Schaefer, P. 1992. Qualitative Physics and Neural Networks for Predicting Environmental Change. Technical Papers, ASPRS/ACSM Resource Technology '92, Vol. 4, pp. 37–46.

Scholten, H., and Stillwell, J. 1990. *Geographic Information Systems for Urban and Regional Planning.* Dordrecht, Netherlands: Kluwer Academic Publishers.

Schutzer, D. 1987. Artificial Intelligence: An Applications-Oriented Approach. New York: Van Nostrand Reinhold.

Sequeira, R., Willers, J., and Olson, R. 1996. Validation of a Deterministic Model-based Decision Support System. *AI Applications in Natural Resource Management* 10:25–40.

Shannon, R. (1975). *Systems Simulation: The Art and Science.* Englewood Cliffs, N.J.: Prentice-Hall.

Simon, H. 1960. The New Science of Management Decisions. New York: Harper and Row.

Simon, H., and Newell, A. 1982. Human Problem Solving. Englewood Cliffs, N.J.: Prentice-Hall.

Simpson, P. 1990. *Artificial Neural Systems.* New York: Pergamon Press.

Skidmore, A., et al. 1996. An Operational GIS Expert System for Mapping Forest Soils. *Photogrammetric Engineering and Remote Sensing* 62:501–11.

Skidmore, A., Ryan, P., and Warwick, D. 1991. Use of an Expert System to Map Forest Soils from a Geographical Information System. *International Journal of Geographical Information Systems* 5:431–45.

Sombe, L. 1990. *Reasoning Under Incomplete Information in Artificial Intelligence.* New York: Wiley and Sons.

Spillman, R., and Spillman, B. 1987. A Survey of Some Contributions of Fuzzy Sets to Decision Theory. In *Analysis of Fuzzy Information,* ed. J. Bezdeck. Boca Raton, Fla.: CRC Press.

Sprague, R., and Carlson, C. 1982. *Building Effective Decision Support Systems.* Englewood Cliffs, N.J.: Prentice-Hall.

Srinivasan, R., and Engel, B. 1994. A Spatial Decision Support System for Assessing Agricultural Non-Point Source Pollution. *Water Research Bulletin* 30:441–53.

Starr, J., and Estes, J. 1990. Geographic Information Systems: An Introduction. Englewood Cliffs, N.J.: Prentice-Hall.

Stein, A. 1995. Interactive GIS for Environmental Risk Assessment. *International Journal of Geographical Information Systems* 9:509–25.

Stern, P., Young, O., and Druchman, D. 1992. *Global Environmental Change.* Washington, D.C.: National Academy Press.

Suen, C., Grogono, P., and Shinghal, R. 1990. Verifying, Validating, and Measuring the Performance of Expert Systems. *Expert Systems with Applications* 1:93–102.

Sui, D. 1992. A Fuzzy GIS Modeling Approach for Urban Land Evaluation. *Computers, Environment, and Urban Systems* 16:101–15.

Suter, G., Barnthouse, L., and O'Neill, R. 1987. Treatment of Risk in Environmental Impact Assessment. *Environmental Management* 11:295–303.

Takagi, H., Suzuki, N., and Koda, T. 1992. Neural Networks Designed on Approximate Reasoning Architectures. *IEEE Transactions on Neural Networks* 3:752–60.

Taylor, R. 1984. *Behavioral Decision Making.* Glenview, Ill.: Foresman and Company.

Thomson, S. 1996. Decision Support Systems to Interpret Soil-Moisture Sensor Readings for Crop Water Management. *AI Applications in Natural Resource Management* 10:57–66.

Tim, U., and Jolly, R. 1994. Evaluating Agricultural Nonpoint-Source Pollution Using Integrated Geographic Information Systems and Hydrological Water Quality Models. *Journal of Environmental Quality* 23:25–35.

Tomlin, C. 1990. *Geographic Information Systems and Cartographic Modeling.* Englewood Cliffs, N.J.: Prentice-Hall.

Turban, E. 1986. Integrated Expert Systems and Decision Support Systems. *MIS Quarterly* 12:121–36.

———. 1992. *Expert Systems and Applied Artificial Intelligence.* New York: Macmillan Publishing Co.

Turner, B., Kasperson, R., and Meyer, W. 1990. Two Types of Global Environmental Change. *Global Environmental Change* 4:14–22.

Veregin, H. 1989. Error Modeling for Map Overlay. In *Accuracy of Spatial Databases,* ed. M. Goodchild and S. Gopal. London: Taylor and Francis.

———. 1994. Integration of Simulation Modeling and Error Propagation for the Buffer Operation in GIS. *Photogrammetric Engineering and Remote Sensing* 60:427–35.

Veregin, H., and Lanter, D. 1995. Data-Quality Enhancement Techniques in Layer-Based Geographic Information Systems. *Computers, Environment, and Urban Systems* 19:23–36.

Vieu, L., and Martin-Clouaire, R. 1994. Spatial and Qualitative Reasoning for Modeling Physical Processes. *AI Applications in Natural Resource Management* 8:61–74.

Vitek, J., Walsh, S., and Gregory, M. 1984. Accuracy in Geographic Information Systems: An Assessment of Inherent and Operational Errors. Proceedings, PECORA IX Symposium, pp. 296–302.

Vlek, C., Timmermans, D., and Otten, W. 1993. The Idea of Decision Support. In *Computer-Aided Decision Analysis,* ed. S. Nagel. Westport, Conn.: Quorum Books.

Vlek, C., and Wagenaar, W. 1979. Judgment and Decision Under Uncertainty. In *Handbook of Psychonomics, Volume 2,* ed. J. Michon. Amsterdam: North-Holland.

Voogd, H. 1983. *Multicriteria Evaluation for Urban and Regional Planning.* London: Pion, Ltd.

VanVoris, P. et al. 1993. TERRA-Vision: The Integrations of Scientific Analysis into the Decision Making Process. *International Journal of Geographical Information Systems* 7:143–64.

Walsh, M. 1993. Toward Spatial Decision Support Systems. *Journal of Water Resources Planning and Management* 119:158–69.

Walsh, S., Lightfoot, D., and Butler, D. 1987. Recognition and Assessment of Error in Geographic Information Systems. *Photogrammetric Engineering and Remote Sensing* 53: 1423–30.

Wang, F. 1994. The Use of Artificial Neural Networks in a Geographical Information System for Agricultural Land Suitability Assessment. *Environment and Planning A* 26:265–84.

Wang, F., Hall, G., and Subaryono 1990. Fuzzy Information Representation and Processing in Conventional GIS Software. *International Journal of Geographical Information Systems* 4:261–83.

Warwick, C., Mumford, J., and Norton, G. 1993. Environmental Management Expert Systems. *Journal of Environmental Management* 39:251–70.

Wasserman, P. 1989. Neural Computing: Theory and Practice. New York: Van Nostrand Reinhold.

Waterman, D. 1986. A Guide to Expert Systems. New York: Addison-Wesley.

Webster, C. 1993. GIS and the Scientific Inputs to Urban Planning: Part 1-Description. *Environment and Planning B: Planning and Design* 20:709–28.

———. 1994. GIS and the Scientific Inputs to Planning: Part 2-Prediction and Prescription. *Environment and Planning B: Planning and Design* 21:145–57.

Weiss, S., and Kulikowski, C. 1984. *A Practical Guide to Designing Expert Systems.* New York: Rowman and Allenheld.

Wenger, R., and Rong, Y. 1987. Two Fuzzy Set Models for Comprehensive Environmental Decision Making. *Journal of Environmental Management* 25:167–80.

Wessman, C. 1992. Spatial Scales and Global Change. *Annual Review of Ecology and Systematics* 23:175–200.

White, H. 1989. Learning in Artificial Neural Networks: A Statistical Perspective. *Neural Computing* 1:425–64.

Whyte, A., and Burton, I. 1980. *Environmental Risk Assessment, SCOPE 15*. New York: Wiley and Sons.

Wigan, M. 1987. Legal and Ethical Issues in Expert Systems Used in Planning. *Environment and Planning B: Planning and Design* 14:305–21.

Wilson, A. 1981. *Geography and the Environment: Systems Analytical Methods*. New York: Wiley and Sons.

Wilson, I. 1978. Scenarios. In *Handbook of Futures Research*, ed. J. Fowels. New York: Greenwood Press.

Winter, K., and Hewitson, B. 1994. Self-Organizing Maps: Applications to Census Data. In *Neural Nets: Applications in Geography*, ed. B. Hewitson and R. Crane. Dordrecht, Netherlands: Kluwer Academic Press.

Winston, P. 1984. *Artificial Intelligence*. Reading, Mass.: Addison-Wesley.

Winston, P., and Prendergast, K. 1984. *The AI Business*. Cambridge, Mass.: MIT Press.

Wood, D. et al. 1988. TEAMS: A Decision Support System for Multiresource Management. *Journal of Forestry* 86:25–34.

Wood, W., and Frankowski, E. 1990. Verification of Rule-based Expert Systems. *Expert Systems with Applications* 1:317–22.

Woodcock, C., Sham, C., and Shaw, B. 1990. Comments on Selecting a Geographic Information System for Environmental Management. *Environmental Management* 14:307–15.

Worrall, L. 1990. *Geographic Information Systems: Developments and Applications*. London: Belhaven Press.

Wright, G., and Ayton, P. 1994. *Subjective Probability*. New York: Wiley and Sons.

Wright, J. et al. 1993. *Expert Systems in Environmental Planning*. New York: Springer-Verlag.

Yager, R. 1977. Multiple Objective Decision Making Using Fuzzy Sets. *International Journal of Man Machine Studies* 9:475–81.

———. 1982. Measuring Tranquility and Anxiety in Decision Making. *Journal of General Systems* 8:139–44.

Yager, R., and Basson, D. 1975. Decision Making with Fuzzy Sets. *Decision Sciences* 6:590–600.

Yan, W., Shimizu, E., and Nakamura, H. 1991. A Knowledge-based Computer System for Zoning. *Computers, Environment, and Urban Systems* 15:125–40.

Yin, Y., and Xu, X. 1991. Applying Neural Net Technology for Multi-Objective Land Use Planning. *Journal of Environmental Management* 32:349–56.

Zadeh, L. 1965. Fuzzy Sets. *Information, and Control* 8:338–53.

———. 1978. Fuzzy Sets as a Basis for a Theory of Possibility. *Fuzzy Sets and Systems* 1:3–28.

———. 1988. Fuzzy Logic. *Computer* 21:83–92.

———. 1994. Fuzzy Logic, Neural Networks, and Soft Computing. *Communications of the ACM* 37:77–84.

Zadeh, L., and Kacprzyk, J. 1992. *Fuzzy Logic for the Management of Uncertainty*. New York: Wiley and Sons.

Zechhaser, R., and Viscuss, W. 1990. Risk Within Reason. *Science* 248:559–64.

Zeleny, M. 1982. *Multiple Criteria Decision Making*. New York: McGraw-Hill.

Zimmerman, H. 1984. *Fuzzy Sets and Decision Analysis*. Amsterdam: North-Holland.

———. 1987. *Fuzzy Sets, Decision Making, and Expert Systems*. Boston: Kluwer Academic Press.

Zonnefeld, J. 1983. Some Basic Notions in Geographical Synthesis. *GeoJournal* 7:121–29.

Index

A

H

L

M

N

O

P

Q

R

S

W

X

Z